CHILDHOOD'S DEADLY SCOURGE

CHILDHOOD'S DEADLY SCOURGE

The Campaign to Control Diphtheria in New York City, 1880–1930

EVELYNN MAXINE HAMMONDS

The Johns Hopkins University Press
Baltimore and London

This book has been brought to publication with the generous
assistance of the Program in Science, Technology, and Society
at the Massachusetts Institute of Technology.

The Johns Hopkins University Press
2715 North Charles Street
Baltimore, Maryland 21218-4363
www.press.jhu.edu

Portions of chapters 2, 3, and 4 appeared in an earlier version
in Elizabeth Fee and Evelynn M. Hammonds, "Science, Politics,
and the Art of Persuasion: Promoting the New Scientific Medicine in
New York City," in *Hives of Sickness: Public Health and Epidemics
in New York City*, ed. David Rosner (New York: Museum of the
City of New York and Rutgers University Press, 1995), 155–97.

Library of Congress Cataloging-in-Publication Data
will be found at the end of this book.
A catalog record for this book is available
from the British Library.

ISBN 0-8018-7097-6

CONTENTS

ACKNOWLEDGMENTS

I am happy to have the opportunity to thank all the people who helped to make this book possible. First, to my teachers and colleagues in the Department of the History of Science at Harvard University, Barbara Gutmann Rosenkrantz, Everett Mendelsohn, and Allan Brandt. They provided me with an unparalleled educational and intellectual experience during my time at Harvard. Without their patience, demand for intellectual rigor, and continuing support, I could never have made what I hope has been a successful transition from scientist to historian of science.

My work also benefited from conversations with Gerard Fergerson, Michael Fortun, Mark Barrow, Judith Walzer Leavitt, Charles E. Rosenberg, Vanessa Northington Gamble, Elizabeth Fee, Susan Reverby, and Jennifer Terry. I especially want to thank Bert Hansen, who at critical points helped to make this a much better book by generously sharing with me his knowledge of the history of disease in New York City.

I thank Anne Fausto-Sterling, Evelyn Fox Keller, Ruth Hubbard, Helen Longino, and Sandra Harding for their friendship, for their unwavering belief that I could become a historian of science, and for the education they gave me in feminist science studies. I thank my colleagues at the Massachusetts Institute of Technology, M. Roe Smith, Deborah Fitzgerald, Michael M. J. Fischer, and especially Kenneth Manning for his support of my work. I can never begin to acknowledge the support I have received from my friends, Barbara Balliet, Elsa Barkley Brown, Vèvè Clark, Cheryl Clarke, and Saj-Nicole Joni.

I owe a special thanks to Joan Wallach Scott. Her careful and critical readings of this manuscript were invaluable. She made the time I spent at the Institute for Advanced Study one of the intellectual highlights of my life.

This book and the dissertation it grew out of were made possible by the financial support of several institutions. I am immensely grateful for fellowship support from Harvard University, Hampshire College, the Ford Foundation, the National Endowment for the Humanities, and MIT. I also thank Dean Philip Khoury of the School of Humanities at MIT for time to complete this work. The funds from the Class of 1947 Career Development Chair at MIT provided crucial support to finish the book.

No book such as this could exist without the help of numerous archivists. I thank the staffs of the Countway Library at Harvard University, the Schlesinger Library at Radcliffe College, Widener Library at Harvard University, the New York Academy of Medicine, the Municipal Archives of New York City, the New York Public Library, the Rockefeller Archives, Sterling Library at Yale University, the Medical Library at Yale University, the Rhode Island Historical Society, the John Hay Brown Library at Brown University, Metropolitan Life Insurance Company Archives, and the Columbia University Library.

I thank Jacqueline Wehmueller, my editor, and the anonymous reader of my manuscript for the timely suggestions that significantly improved the manuscript in important ways.

Finally, I thank my family, William E. Hammonds, Kathy Hammonds-Slaughter, and Cathryn Camp for their love and support. A very special thanks to Alexandra E. Shields for her editorial help on this book and for all that she has brought to my life.

This book is dedicated to my mother, Evelyn Baker Hammonds (1930–1985). Just before I began my studies at Harvard, my mother was diagnosed with ovarian cancer. In the midst of her pain, in the face of her impending death, she gave me her support and her faith that I could do well in my new career. She fought her illness as she lived her life, with true dignity and courage. She died on November 2, 1985, three days before her fifty-fifth birthday. From that day to the completion of this book, I have relied on her spiritual presence to see me through this process. I could not have written one word of this book without her support. The memories of my childhood days with my mother, my first teacher, my best teacher, who lovingly conveyed to me the lessons I draw on every day, have sustained me. I have tried to make this book a work that would meet her rigorous standards. It is a small token of gratitude for all that she gave to me.

CHILDHOOD'S DEADLY SCOURGE

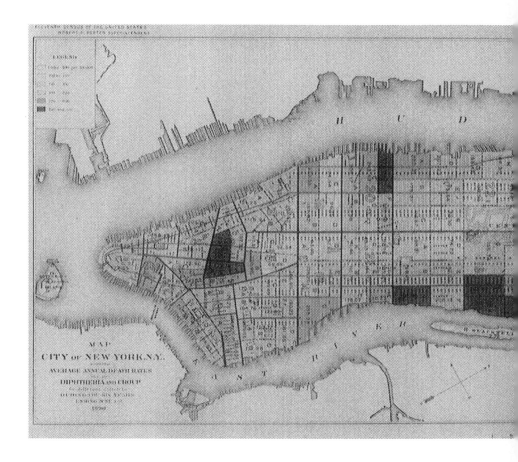

Map of the City of New York, N.Y., Showing the Average Annual Death Rates Due to Diphtheria and Croup in the Different Districts during the Six Years Ending June 1, 1890.

SOURCE: *John S. Billings,* Vital Statistics of New York and Brooklyn: Covering a Period of Six Years Ending May 31, 1890 *(Washington, D.C.: Government Printing Office, 1894).* NOTE: *The death rate in 1890 was 123.44 (per 100,000) and the number of recorded deaths was 1,870, with the highest number of deaths being among children under five years old. The map shows the relative prevalence of diphtheria and croup in different parts of the city. The darkest shaded areas indicate that the districts where diphtheria was especially fatal were mostly low-lying areas along the banks of the East River and the Hudson River. Diphtheria was also prevalent in the most densely populated districts in lower Manhattan, with some exceptions in less populated districts near Central Park.*

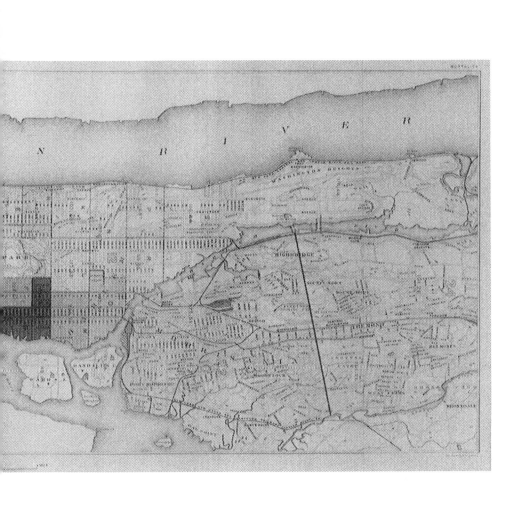

INTRODUCTION

Early in 1993 reports began to appear in the U.S. press of a diphtheria outbreak in Russia and other regions of the former Soviet Union.[1] The outbreak in Moscow began in 1990 when a few infected paramilitary recruits arrived in the city from southern Russia where diphtheria is commonplace but causes only a mild skin disease. Moscow residents, however, were experiencing the more severe forms of the disease. As the state-controlled health care system collapsed, the disease spread quickly across the region. By 1994, 48,000 cases of diphtheria had developed. One year later there were more than 80,000 cases, with 2,000 deaths throughout the former Soviet Union.[2] That year a group of public health experts from WHO (World Health Organization) traveled to Russia to survey the situation. The outbreak had begun to spread across the borders of Russia into Western Europe. "We're worried and believe me, the Russians and their new neighbors are worried," wrote one of the observers.[3] While Western observers characterized the Russian situation as an "accident waiting to happen" because so few Russian children had received the full series of diphtheria immunizations, fear was also expressed about the high levels of susceptibility to the disease in the adult population in Western countries. The Centers for Disease Control (CDC) estimated that 20 to 60 percent of the adult population over the age of twenty in Europe and the United States were susceptible to diphtheria.[4]

The news from Russia was alarming. Not since the 1930s had severe epidemics of diphtheria occurred in this country. Only forty cases of diphtheria were reported in the United States between 1980 and 1993. Given the disease's long decline, the CDC had set the goal of its complete elimination by 1996.[5] The CDC became alarmed about the news from Russia because the immunization status of the majority of adults in the U.S. population is unknown. To be completely effective, booster vaccinations of toxoid are needed approximately every ten years to maintain a high level of immunity against diphtheria in adult populations. Yet there had been no systematic attempt to maintain immunity in adult populations in the United States, even though virulent strains of diphtheria still appeared in many regions of the world.[6] With a breakdown in government-supported immunization programs such as that in the former Soviet Union, and an increase in migration and travel from regions where virulent strains still existed, the potential for the spread of diphtheria has increased.

Yet the reemergence of diphtheria epidemics cannot be completely explained by gaps in immunization programs. Diphtheria's reemergence, along with that of other infectious diseases such as tuberculosis, and the pandemic of acquired immunodeficiency syndrome (AIDS) came at a time of great complacency regarding the danger from infectious diseases.[7] Until the advent of AIDS in particular, many people viewed infectious diseases simply as remnants of our premodern past. This complacency was supported by the widely accepted narrative depicting the triumph of scientific medicine in bringing these diseases under control. In its simplest form, this narrative asserted that the introduction of the germ theory, followed by discoveries in bacteriology and immunology, had produced more accurate ways of identifying the bacteria associated with these diseases. This in turn led to the development of new treatments, culminating in the invention of drugs for prevention. This narrative, privileging the role of science while downplaying other factors, fostered and contributed to the belief that scientific knowledge alone led to the control of infectious diseases that had once ravaged human populations.

This book challenges such a narrative about the control of infectious diseases by examining the history of the control of diphtheria in New York City. Diphtheria was the first infectious disease to be controlled by advances in scientific medicine, particularly discoveries in bacteriology and immunology. The research in bacteriology that led to the development of diphtheria antitoxin was awarded the first Nobel Prize in Medicine in 1901. By the 1930s it had become the paradigmatic disease of the so-called bacteriological revolution and the symbol of the triumph of scientific medicine in the control of infectious disease. New York City is the site of this study because its health department was the first in the United States to apply bacteriology to the control of diphtheria. It established the first bacteriological diagnosis program in 1893, the first public production and distribution of diphtheria antitoxin for treatment in 1895, and the first citywide use of toxin-antitoxin for the active immunization of hundreds of thousands of children by 1930. In addition to its pioneering public health programs, the New York City Health Department's research laboratory made fundamental contributions to the understanding of the etiology of diphtheria—its bacteriology, epidemiology, and immunology.

A number of questions about *how* the successful control of diphtheria was achieved are addressed in this book. How were the interventions of bacteriological diagnosis—diphtheria antitoxin and toxin-antitoxin immunizations—implemented by bacteriologists, public health experts, and physicians under the auspices of a municipal health department? How did these interventions affect medical and public health theory and practice about diphtheria and by extension other infectious diseases? What role was played by professional politics between advocates of bacteriology, physicians, and public health experts in the acceptance of these interventions? How was the public encouraged to accept them? I pay close attention throughout to the changing conception of diphtheria engendered by these laboratory interventions and to the questions about this disease that remained unanswered despite successes that led to its control.

In the end, I contend that diphtheria was not controlled in any direct or straightforward way by the introduction of bacteriological knowledge—it was not an instance of simply the right knowledge producing the right outcome. Nor was it the result of brilliant political machinations by public health experts. Instead, it was a process influenced not only by science and politics but also by the nature of the disease itself. The specific biological characteristics of diphtheria, the fact that the bacillus was found in the healthy, its modes of transmission, and its prevalence in New York City all shaped public health policies, therapeutic options, and public perceptions. Diphtheria made some practices and politics possible, while it constrained others.[8]

The successful control of diphtheria in New York City cannot be explained either by a story of the inevitability of scientific progress or by positing that scientific knowledge prevailed only by dint of extraordinary politics. Diphtheria was controlled by two interrelated factors: first, the interaction between the professional, factional, and political interests of those who sponsored, enabled, and resisted the application of bacteriology to medicine and public health.[9] Second, it was controlled by the real scientific advances produced by its transformation in the laboratory and the translation of those transformations into effective practices.[10] Most important, the campaigns to implement these advances involved the reproduction and dissemination of a teleological narrative, so that each advance—bacteriological diagnosis, antitoxin, and immunization—became retrospectively the natural and necessary solution to the problem of diphtheria. Yet each success and advance in knowledge of diphtheria simultaneously revealed how incompletely it was known. Still, by 1930 the story of the triumph of science over diphtheria dominated all accounts of the history of this disease.

Diphtheria was one of the most prevalent diseases in New York City in the late nineteenth and early twentieth centuries. The mortality rate was frighteningly high, with children being its most frequent victims. Unlike cholera, smallpox, or yellow fever, diphtheria did not manifest itself in spectacular epidemics, leaving thousands dead in a short period of time. Despite yearly fluctuations, diphtheria was endemic and took a steady toll, claiming more than a thousand deaths per year in the 1890s.

At the time, diphtheria was known as an infectious and contagious disease that mainly affected young children. In milder forms of the disease, victims had a simple sore throat, while in severe cases victims either suffocated from the buildup of a pseudomembrane that blocked the throat passages or died from the effects of a toxin produced by the bacteria.[11] As with many other infectious diseases in the nineteenth century, diphtheria was generally identified by its symptoms.

Until the 1890s diphtheria was a difficult disease to diagnose and to treat, and no cure was available. More important, it was difficult to recognize because its symptoms were similar to other inflammations of the throat. Diphtheria's uncertain identity continued into the bacteriological era despite the identification of the bacteria that caused it. The bacteriological diagnostic program initiated in 1894 made it possible to chart every reported diphtheria case onto a map of the city. For the first time, the threat posed by diphtheria could be represented visually; not only individual houses and apartments but also entire city blocks could be monitored on a visual field.

Once diphtheria's incidence was made visible, its impact in the poor and largely immigrant districts of the city was evident. The continuing unabated presence of diphtheria in these districts despite the new scientific advances seemed to support the need for the expansion of municipal public health activities. Elite groups considered the intervention into the homes of the immigrant poor an appropriate response, especially given their own fears of contamination from these groups. Yet when these poor died from diphtheria, or when the number of cases continued to rise despite scientific advances, city health authorities blamed physicians' resistance to the newest scientific advances and the health department's authority, as well as parental neglect. Thus the introduction of bacteriological-based methods to control diphtheria was affected by political processes that had no direct relationship to science or medicine.

Bacteriology triumphed as the right solution to the diphtheria problem because city health authorities used it as a vehicle for establishing the power of laboratory science in the control of disease and mounted campaigns to demonstrate that they were right. Although some historians have characterized physician opposition or resistance to these interventions and campaigns as opposition to bacteriology and laboratory science, this misses the point that belief in bacteriology had institutional and professional consequences. Public resistance to the health department's interventions was overcome by assuring residents that the autocratic police powers of the health department would be exercised only when the health of the community was threatened and by emphasizing that interventions were based, not on political grounds, but on objective scientific knowledge. In the case of diphtheria, the interventions worked, thus further justifying the notion of interventions in the name of science.

The competition for status and authority between public health experts and the medical community is a continuing theme in the history of the control of diphtheria. The New York City medical community in the late 1890s was fluid, chaotic, and riddled with factions. In the midst of an economic depression, general practitioners trying to establish private practices increasingly feared the growing coalition between specialists, medical college faculty, and experts in public health.[12] The boundary between the interests of private curative medicine and public preventive medicine was constantly being redrawn. In a city dominated by the politics of the Tammany Hall machine, known for its widespread use of public jobs for political patronage, public health professionals were viewed with skepticism by the public and by medical leaders. Those who sought to ground the professional authority of public health and preventive medicine, as John Harley Warner has argued, "had first to come to terms with the underlying shift in professional identity it implied, from a primary rooting in practice to one in special knowledge. And this involved fundamental alterations in their understanding of the relationship between science and professional identity, between science and professional morality."[13]

The men who developed and implemented the diphtheria control programs made the strongest possible link between practice and science. They used this link to make social and political conflicts more susceptible to purely technical solutions. These men successfully translated their bacteriological-based research into a viable and influential public health program. They rhetorically and practically deployed science in such a way that "science not only provided the answers; it also defined the problems and delimited them in such a way that only science could solve them."[14]

In addition to the rhetorical link between knowledge and practice, these public health leaders also recognized the need for an institutional location in which to produce scientific knowledge that had somewhat flexible boundaries. In two instances—first when the tools for bacteriological diagnosis were made available to physicians in drugstores, and later when immunizations were given in public parks—the laboratory was extended beyond the confines of its physical location to become a visible symbol of scientific medicine. The expanded conception of the laboratory was made possible in New York City because it was part of a public institution.

A final theme of this book addresses new claims to authority made by the American medical profession in the name of laboratory science during the Progressive Era. Sociologist Paul Starr has argued that "as the American faith in democratic simplicity and common sense yielded to a celebration of science and efficiency, doctors were able to claim a new measure of cultural authority over a public impressed by the achievements of the laboratory and ready to accept that it was complex beyond their mastery."[15] Neither Starr nor his critics have adequately explained how this new cultural authority of physicians and scientific medicine was established. In the absence of such analysis, it is often concluded that the reason lay in the objective power of science or medical knowledge itself. Yet in order to accrue cultural authority, scientific medicine had to be comprehensible and persuasive to socially powerful groups and the lay public. Public health leaders in New York City saw the antitoxin subscription and immunization campaigns as critical to the success of their diphtheria-control programs. The success of these programs in turn secured the authority of science more generally, even though other diseases did not, in fact, yield as easily or completely to laboratory and medical intervention. The campaigns against diphtheria, then, were crucial moments when medicine's new cultural authority was constructed and represented.[16]

This book is organized around four aspects of the diphtheria-control program in New York City: bacteriological diagnosis, the introduction of antitoxin, the carrier problem, and active immunization. With the exception of the carrier problem, these topics have typically been identified in retrospective accounts as the critical interventions that brought about control of this disease. I begin in Chapter 1 with a discussion of the identification and naming of diphtheria in the early nineteenth century prior to the introduction of bacteriology, in order to illuminate how it was transformed by the laboratory at the end of the century.

Bacteriological Diagnosis and the Introduction of Antitoxin

In 1893, the New York City Health Department made bacteriological diagnosis of diphtheria mandatory for all suspected cases of the disease in the city. For the next three years, following the introduction of diphtheria antitoxin in 1895, physicians and advocates of bacteriology fiercely debated the implications of these new methods of diagnosis and treatment. Physicians, with a deep belief in the "epistemology of experience" as a guide to clinical practice, expressed their skepticism toward bacteriology as a science and the health department as an institution. Clinical and bacteriological knowledge came into conflict as physicians questioned the bacteriological conception of diphtheria in the process of evaluating the antitoxin for treatment. At issue for physicians were the financial stakes in protecting themselves from the intrusions of health inspectors and health department officials as well as the threat that bacteriological diagnosis and antitoxin posed to their professional control over diagnosis and treatment of disease. The campaign to promote diphtheria antitoxin among the public threatened to undermine their public authority as well.

In the opposing corner are the proponents of bacteriology, identifiable as a professional group of public health experts at this point only because of their support of bacteriology and their institutional location in the health department. For this group, *control* of disease was the preeminent goal. Control, as Hermann Biggs[17] and William H. Park[18] came to define it, required the efficiency and certainty that bacteriology provided. Park's innovative "culture kit" became the interface between the laboratory and the bedside. Biggs's diagnostic program, organized around the culture kit, gave physicians direct access to bacteriological diagnosis. Thus Park's kit made bacteriology visible, while Biggs's practices made it accessible. It was this accessibility and visibility that made the dispute over the meaning of this new laboratory-based knowledge so contentious in the New York City context. The diagnostic program, followed within months by the introduction of antitoxin, won allies for the new public health programs while also providing increased opportunities for resistance on the part of physicians and parents. The new public health experts were allied with the medical profession in their belief that public and private health decisions should remain under medical supervision. They differed from the profession, however, in their perception of the boundary between public and private medicine. This difference would be even more apparent in the 1920s when the health department tried to marshal physicians' support for carrier control and immunization. While bacteriological knowledge appeared to provide a sure ground on which to base public health practice in the 1890s, this assurance was destabilized by the early decades of the twentieth century.

The Carrier Problem

Public health emerged as a more visible specialty in the New York medical community by the first decades of the twentieth century. During this period the knowledge that grounded public health practice was in transition. Less attention was paid to sanitary reforms such as garbage removal and street cleaning, and more was given to the control of contagious diseases. There was increasing reliance on science to provide the means to identify, cure, and prevent disease without incurring the costs of major social changes that the earlier focus on the environment required. This faith in science was unsettled when bacteriological knowledge about diphtheria came into conflict with epidemiological knowledge, causing dissension among leading public health experts over how to control the healthy carrier of the disease.

The problem was a complex one. The architects of the diphtheria-control programs had built support for their regulations and interventions among physicians and the public by claiming that bacteriology would "rid diphtheria of its terrors" and lead to its eradication, yet diphtheria remained imperfectly controlled. Many cases of the disease were not diagnosed on the first day or two when the antitoxin was most effective, and thus uncontrolled outbreaks continued. Epidemiological data showed that healthy carriers of diphtheria were the source of outbreaks. The bacteriological data confirmed this. It thus seemed self-evident that if diphtheria were to be controlled, healthy carriers had to be controlled. Yet these people were not sick, as everyone was quick to note.

The problem that carriers presented was enormous. One to two percent of the child population of New York City were carriers on any given day. Epidemiological data showed that a high percentage of diphtheria cases were infected by carriers. Given the incidence of the disease in the city, thousands of carriers were presumed to be transmitters of infection. The epidemiological data could only confirm that carriers were prevalent. Without laboratory tests to determine the virulence of the bacilli found in the healthy carriers, there was no way to measure their actual danger to others.[19] The laboratory tests to confirm the virulence of bacteria in healthy carriers were time-consuming and costly. The carrier dilemma threatened to undermine the effectiveness of the health department's diphtheria program.

Clinical, epidemiological, and bacteriological knowledge about diphtheria was needed to determine the "danger" from carriers. Public health experts could not convince physicians or the public to submit to multiple throat cultures, quarantine, isolation, and constant inspection and removal of children from schools in order to expose and isolate carriers. The implementation of these practices in public schools was especially problematic. In the 1920s the public schools were increasingly sites of conflict between immigrant groups and health officials in New York City.[20] Parents were highly suspicious of actions taken by health officers directed toward their children's health. Health officers, on the other hand, perceived parental fears as unfounded resistance to scientific medicine. Public health experts were caught between three groups: physicians allied with their patients in asserting that the health department should provide health education, not health care; parents wanting to protect their children; and laboratory scientists, emphasizing the danger from carriers. They could not adopt the drastic measures that science demanded to control diphtheria carriers. Science had paradoxically produced unexpected limits to public health practice while it simultaneously provided the impetus for more sophisticated epidemiological studies of the carrier.

Active Immunization

The problem of high diphtheria morbidity became a regular theme of public health experts in the 1920s. Diphtheria was the disease in which scientific knowledge was the most complete, and thus they expected that it should have been under control. That it was not was a medical, political, and public health problem. In this period William Park saw the fruition of his long efforts to extend the immunity produced by diphtheria antitoxin through the development of active immunization with toxin-antitoxin. Immunization was an intervention that was simple, relatively inexpensive, and had fewer social costs than any carrier-control program. Its implementation, however, required the cooperation of many groups in the city.

Initially, immunization efforts were directed toward school-age children. The group most susceptible to diphtheria were children under the age of five. The campaign was developed because public health leaders faced the problem of how to reach pre-school-age children. Repeated entreaties to physicians to voluntarily take up immunization of this group yielded scanty results. The full cooperation of physicians was essential because immunization had to be complete throughout the city to be effective.

Physicians and public health experts were more clearly divided in their economic and professional interests by the late 1920s. An argument for immunization based solely on its scientific merits did little to stir physicians to action. They were a much more highly organized profession by this time. The American Medical Association (AMA) was ever alert to public programs that threatened physicians' economic interests. They had won several notable battles with cities, state governments, and private philanthropies during the 1920s over health programs directed at the poor, most notably the defeat of the Shepherd-Towner Act.[21] As with the introduction of antitoxin, the health department turned to a highly visible public campaign to promote immunization.

Using every form of media available, public health leaders made immunization a "public" issue. In so doing, they were able to override physicians' resistance with a campaign that galvanized the public. The campaign both chastised private physicians for resisting public health measures and rewarded their participation by subsidizing the costs of immunization. The goals of preventive medicine, however, would not be subordinated to private medical interests. The rhetoric of this campaign wrote the history of diphtheria control in New York City in terms that positioned science, public health experts, and physicians, in that order, as the ultimate child savers. Despite their clearly lesser position, physicians were as engaged by the message of the campaign as was the public. Who, after all, did not want to be allied with saving children's lives?

The programs to control diphtheria were representative of the ethos that characterized many Progressive Era reform efforts. Hermann Biggs and his successors in the New York City Health Department practiced a kind of science-based paternalism that espoused that public health should dominate private medicine in matters of disease control. They downplayed the inevitable tensions that their practices engendered between private and public medicine, technical control and social order, expert knowledge and egalitarian democratic ideals, and the needs of the poor versus their rights as citizens. For Biggs the goals of public health overrode such concerns, as he argued in one of his more well known public addresses: "Sanitary authorities must protect the community from the individual, in order to provide the greatest good to the greatest number."[22] Indeed, in Biggs's terms, "the State, not the individual, would define the common goal and see to its fulfillment." He was committed to the view that "only the enlarged authority of the government could satisfy the particular needs of all citizens."[23] This absolute faith in the power of the state to identify the common good and produce positive outcomes, among which I count the interventions to control diphtheria, masked the problems with this view. Biggs in particular, but also his successors, failed to see that the control of diphtheria would always be an ongoing process, in part because it was implemented by government agencies constrained by politics and also because the scientific knowledge was incomplete. They believed science could bring about the perfect control of diphtheria by producing the technology to eradicate the disease. Science would also give them the authority to contain the political process by turning control of disease into a moral good that transcended politics and private interests. The legacy of Biggs's programs was that the absolute control of diphtheria remains an elusive goal. Still the narrative of scientific success engendered by the control of diphtheria was used as a justification for programs much broader and ultimately less successful than this one, establishing for scientific medicine an authority beyond its own limits.

A short note on the current conception of diphtheria may provide a helpful context for the reader. In modern terms, diphtheria is defined as an acute infectious disease caused by the bacillus *Corynebacterium diphtheriae*. This bacillus occurs in three major types: *gravis* and *intermedius*, both of which tend to be associated with high case fatality rates, and *mitis*, which is associated with milder infections.[24] The bacillus usually enters the body through one of three avenues: the upper respiratory tract, mainly via droplets or discharges from infected persons; the secretions from the nose and throat of convalescent cases; or from the healthy throats of individuals who acquired the bacilli from being in contact with others having virulent germs. Clinical manifestations occur one to seven days after infection with one or more strains of the bacillus. The disease is characterized by a local inflammatory lesion, usually in the larynx, called a "pseudomembrane,"[25] and a toxic reaction involving primarily the heart and peripheral nerves. Death can result from respiratory obstruction by the membrane or from effects of the toxin on the heart, nervous system, or other organs.[26] Although diphtheria toxin plays a part in the formation of the membrane, its systemic effects following absorption are by far the most important, and diphtheria, like tetanus, is essentially a toxemia. The toxin can cause severe damage to kidney, heart, and nerves.

1: THE IMPOSSIBILITY OF CONTROL

It takes the lofty as well as the lowly, it spares neither the rich nor the poor; its rav-
ages are witnessed at the sea-side and in the crowded city, on the mountain and in
the valley. And whenever an epidemic visits us, although with our present knowl-
edge we may not be able to "stamp it out" as we do with small-pox by vaccination,
we may, by attention to sanitary measures, deprive it of its greatest power, and thus
remove the aggravating if we cannot control the primary cause.

NEW YORK CITY HEALTH DEPARTMENT, 1870S

If, as Charles Rosenberg asserts, "disease does not exist in our culture as a
social phenomenon until we agree it does — until it is named," then any study
of disease must begin with the naming process.[1] What is now called diphtheria was known under a variety of names until the 1820s, including morbus
stranglatorius, croup, malignant ulcerous sore throat, angina maligna, and
throat distemper. Although the name *diphtheria* was coined in 1826, its identity was contested in medicine and public health until the end of the century. This chapter examines physicians' and sanitarians' efforts to define
diphtheria up until the introduction of bacteriology.

Diphtheria meant many things to physicians, depending on whether they
studied the disease in the clinic, in the homes of their patients, or in bodies
at autopsy. The clinical picture of diphtheria was a complicated one. The
presence, absence, and character of the pseudomembrane and the various
symptoms associated with it drew the greatest attention. Yet the presence of
a pseudomembrane did not always signal the presence of diphtheria, nor did
its absence in the presence of other symptoms indicate that a person did not
have diphtheria. Thus, even after diphtheria was named as a disease entity, it
was routinely confused with other ailments affecting the throat. The question of whether diphtheria was one disease or several was not completely settled by any of the prevailing theories of disease.

Those studying epidemics and the prevalence of diphtheria in populations were hindered by the problem of definition as well as by inadequate explanations of how it was spread. At the beginning of the nineteenth century epidemics of infectious sore throat appeared throughout Europe and the United States. Outbreaks continued in sporadic waves with increasing virulence until the 1850s in the United States. By the end of the century it was endemic in most urban areas on the East Coast and still epidemic in rural areas across the country. The outbreaks at the end of the century were sometimes mild and other times severe, and no single theory could explain their changing character. To give an accurate account of the prevalence of the disease before the 1850s is impossible because of gaps in reporting and because various names were used to describe the disease. Deaths from diphtheria first appear in the New York City Department of Health records in 1857 along with deaths from other throat inflammations. There was a slow increase in the number of deaths attributed to diphtheria over the years as reported deaths from other throat diseases declined.

Each generation of physicians, researchers, and sanitarians that encountered diphtheria noted the problems in the clinical and epidemiological explanations of it. It is not surprising, then, that the discourse about diphtheria is permeated with language about control—or rather the lack of control— that these experts had over the disease. If theories or explanations of disease and the process of naming disease reflect attempts to control the fear and uncertainty that the presence of disease produces, then the persistent expressions of lack of control over diphtheria reveal the inadequacy of each explanatory framework applied to it.

The Identity of Diphtheria

The various names attributed to infectious inflammations of the throat in the health department records point to the major question in the study of diphtheria in the prebacteriological period. From the beginning of the modern history of diphtheria in the early nineteenth century, the study of this disease was plagued by problems in its definition. The first breakthrough in research on the disease that became known as diphtheria occurred in 1826 with the work of the French physician Pierre Fidèle Bretonneau (1778–1862). As chief of medicine at the Hospice Generale at Tours, Bretonneau first reported on diphtheria in 1821. This work, regarded by both his contem poraries and historians as a classic, was the first to describe diphtheria as an etiological and clinical entity.[2] While stationed at a military garrison at Tours, Bretonneau performed careful autopsies on men who had died during an outbreak of *angina maligna*. Comparing these results with observations of three epidemics of malignant sore throat that had occurred in 1818, 1825, and 1826, he concluded that he was observing one disease, not several. The characteristic mark of the disease, he observed, was a pseudomembrane in the pharynx, larynx, or both.[3] He named the disease *diphthérite* (from the Greek word meaning inflammation of a skin, hide, or membrane).[4] Bretonneau defined diphthérite (called diphtérie by 1855) as a single, specific, localized, contagious disease characterized by a false membrane produced by the action of an unknown virus.[5] A leader in the Paris clinical school of medicine, Bretonneau was one of the few who supported the doctrine of specific disease and contagion.[6] Diphtheria emerged as a distinct clinical entity through Bretonneau's association of the specific character of its inflammation and associated lesion, which he argued was produced by a specific (although undetermined) causal agent.

Bretonneau's diphtheria was an exemplar of the Paris schools' method of establishing definite clinical entities by ascribing to each disease a typical course and sufficiently well marked characteristics to justify the designation as a distinct entity.[7] Fairly quickly it was noted that there were problems with Bretonneau's definition, especially its critical dependence on the specific anatomical characteristics of the pseudomembrane. Physicians had long observed that other maladies—scarlet fever, measles, and most notably "croup"—were also associated with the production of a false membrane.[8] Croup was most often compared with malignant sore throat in the early nineteenth century. This disease was as ubiquitous as its name, which referred to a variety of mild and severe inflammations of the throat that often involved obstruction of the windpipe. Bretonneau's postmortem examinations showed that the membrane in croup exhibited the same structure as that in malignant angina, now diphtheria, whereas it had nothing in common with that seen in scarlet fever. In his view, therefore, croup and diphtheria were one and the same.

Although his definition of diphtheria was accepted among many physicians in France, his views were less popular in Germany and England. Several questions were raised by a range of observers. German researchers in pathological anatomy questioned Bretonneau's assertion that the local lesion in diphtheria was the manifestation of a general disease; the British disagreed with the inclusion of croup in the definition of diphtheria. On this last point, even some French researchers concurred that the differentiation of croup from diphtheria was an unsolved, if not an actually insoluble, clinical problem.[9]

The work of Rudolf Virchow and his followers directly challenged Bretonneau's definition of diphtheria as a specific disease. In the 1840s, Virchow's school of cellular pathology began to occupy a prominent place in experimental investigations of disease processes.[10] He stressed the various forms of inflammation of the mucous membrane, and he observed no difference in the form of membranes found in patients who had died of Bretonneau's diphtheria from those found in cases of scarlet fever, cholera, or dysentery. He thus argued that there was no reason to suppose that the croupous or pseudomembranous deposits found in diphtheria were defining characteristics of that disease.[11] Virchow rejected the French teaching that regarded diphtheria and croup as mutually interdependent, forming one specific disease. Instead, he drew a sharp line between croup and diphtheria.[12]

The debate about the form and structure of the pseudomembrane seen in diphtheria and other infectious diseases was not an idle one. The confusion among clinicians and researchers was extensive. For Virchow, diphtheria was a particular cellular reaction.[13] He and his students used the terms *croup* and *diphtheria* in an adjectival sense to designate morphological differences they observed in pseudomembranes examined in autopsies. These terms were used without reference to specific diseases.[14] They named all cases of pseudomembranous inflammations of the air passages—even those due to "taking cold," inhalation of an irritating vapor, or laryngitis—as diphtheritic.[15]

Some practitioners took a different tack, focusing on the effects of the inflammation rather than the lesion. As one physician noted, "Practitioners, on the other hand, apply the term diphtheritic for all inflammations which occur as local manifestations of the specific disease known as diphtheria, and to such inflammations only, whatever may be their form."[16] Other physicians began to replace the clinical conception of diphtheria with an anatomical one. All manner of diseases that were characterized by the presence of some form of pseudomembrane were described as "diphtheritic".[17] In England, and certainly in the United States, most physicians had little access to the pathological-anatomical work being carried out on the Continent. The British in particular held that there was no anatomical basis for the separation of the diphtheritic and croupous process. When a differentiation was attempted, it was based purely on clinical observations.[18]

Physicians continued to debate the nature of croup and diphtheria. The link between diphtheria and croup was a strong one. Diphtheria could not be defined without defining croup.[19] The lack of specificity in the definition of croup was carried over into the definition of diphtheria. Some French writers spoke of "diphtheritic croup" as contrasted with "true croup," while simple laryngitis began to be called "false croup."[20] This language only underscored the indeterminacy in the definition of both diseases.

The debates then turned from the meaning and character of the pseudomembrane in diphtheria and croup to the source and extent of their contagiousness. While some argued that diphtheria was contagious, others argued that croup was not. But if diphtheria was indeed a contagious disease, physicians surmised, then cases of it should produce only other cases of diphtheria, not croup or some other ailment. As one physician put it, "A process can be called contagious only if it can reproduce the same process by transmission to others. If diphtheria were so absolutely different from croup, it always ought to produce diphtheria and not occasionally catarrhal or croupous affections too."[21]

This definition of a contagious disease was based on the model of smallpox, which was known by a well-defined set of symptoms and affected each person only once. The course of the disease was well known, especially the local production of pustules on the skin.[22] Diphtheria did not fit this model. It was apt to recur in those who had once had it, and its symptoms varied. Part of the problem with diphtheria was that both mild and severe forms of it were seen. Although the onset of diphtheria was very similar to croup, in severe cases it showed marked differences. Bretonneau had defined diphtheria as it appeared in its most severe form, leaving the question of the recognition of it in its earliest stages or in mild forms unresolved. But it was precisely at the earliest stages that diphtheria was hardest to recognize. The physician needed years of experience before he could easily distinguish a case of diphtheria among the various forms of sore throats with or without membranes.

The questions of the identity of diphtheria, its relation to croup, and whether it was a local or constitutional disease remained highly contested through the 1880s. Physicians relentlessly described and examined the membrane. As Abraham Jacobi, the leading writer on diphtheria in the United States wrote, "This membrane has been examined and described a thousand times, and its general condition, whether thick or thin, small or large, white or grayish or brownish . . . all this is a matter of degree only, not of kind."[23] Other researchers continued to produce descriptions of cases to support or contest the view that the disease was localized in the throat or was a more general affliction.

As outbreaks of diphtheria increased in the mid-nineteenth century, questions about the causal agent and its mode of transmission came to the fore. In the 1850s British sanitarians defined diphtheria as a *zymotic* disease. Zymotic diseases, or filth diseases, were propagated "either by inoculation and (contact) or by inhalation (infection), in which there was a propagation of the morbid principle."[24] According to William Farr, the British sanitarian who coined this term, zymotic disease was the result of specific poisons of organic origins, either derived from without or generated within the body. For each distinct disease, he posited the existence of an "exciter" that affected particular organs more extensively and more frequently than others and thus gave rise to "specific pathological formations or secretions" in the body.[25] Zymotic diseases were transmitted through the atmosphere by particles that floated in the air, forming a "morbid atmosphere." The density of this atmosphere was in proportion to the proximity of the bodies from which it emanated. Farr based his theory of zymotic diseases on what was known about smallpox. Reasoning by analogy, he added other contagious diseases to this category. In the case of diphtheria, a number of problems arose with such a classification.

The attempt to make diphtheria fit neatly into the category of zymotic diseases was more a testimony to the power and authority of Farr's nosology than to the specific facts of diphtheria. Three questions about the transmission of diphtheria emerged by the middle of the nineteenth century. The first was about the extent to which diphtheria was dependent on unsanitary conditions. The second concerned the unusual cyclical nature of epidemic outbreaks of diphtheria. And the third centered on the nature of the so-called poison that caused diphtheria.

Many observers believed that overcrowding provided very favorable conditions for diphtheria, especially among children sitting in poorly ventilated schoolrooms or living in tenement houses in neighborhoods with great accumulations of filth. Although some observers would claim that such conditions did not cause diphtheria, they certainly felt that filth contributed to its virulence and intensity. The prevailing view was that poverty and filth were the breeding grounds of zymotic diseases. However, this notion was coun-

tered by other facts. Reports noted that epidemic outbreaks of diphtheria occurred just as often away from "noxious influences." Diphtheria was found
among both the poor and the rich. Some physicians began to suspect that
the search for the cause of diphtheria in filth represented the failure to identify the cause of the disease.[26]

In spite of the evidence that diphtheria was not absolutely dependent on
unsanitary conditions, some observers could not let go of the issue comfort-
ably. In order to explain outbreaks of diphtheria in the homes of the better
classes, the concept of an "artificial atmosphere" was invoked. This atmosphere was ostensibly created during the winter months by the lack of ventilation in heated rooms. It existed in places where people had more contact
with each other, thus providing increased opportunities for the transmission
of the diphtheritic poison. As a result, the better classes were made as susceptible to the disease as the lower. Though this explanation had its supporters,
the question of whether unsanitary conditions or the lack of ventilation were
key to the propagation of diphtheria was not settled.

As to the second question, regarding the cyclical nature of the outbreaks,
there was further uncertainty. Though the few available statistics on mortality were extremely unreliable, they tended to confirm this view.[27] The cycles
extended over periods of various lengths, some of a few years and others as
long as a decade. Earlier in the century, the reappearance of diphtheria after
a dormant period led some physicians to believe that they were seeing outbreaks of a new disease. Physicians also noted that outbreaks were not of the
same intensity. Sometimes they were mild with few deaths; at other times
they were severe. Few explanations were offered as to why the disease be-
haved in this way, but the changing character of the outbreaks was a crucial
but puzzling aspect that had implications for the treatment of patients and
the control of outbreaks.

Third, the question of the nature of the "virus" that caused diphtheria increasingly occupied the attention of physicians and sanitarians. Most observers agreed that the disease was contagious, though the nature and mode of
transmission of the infectious agent was not known. By 1880, as some in the
medical world hoped that laboratory research on microorganisms would settle the issue of the cause of contagious diseases, others held to the view that
the diphtheria was caused by "the atmosphere, in which the poison of the
disease is suspended."[28]

This retreat to the atmosphere as the ultimate cause of diphtheria reflected the faults in zymotic theory, as did the characterization of diphtheria

as a zymotic disease on the whole. As Coleman and Cooter have argued, the predominance of the atmosphere as the cause of zymotic disease meant the diminution of other potentially influential relations. In particular it suggested "a decidedly lesser role for, and interest in, the possibility of person-to-person communication of disease; that is, atmospheric predominance renders impotent the contagionist argument."[29] Zymotic theory thus cast diphtheria as a disease that affected all classes equally. The factor of person-to-person transmission of diphtheria was noted by all observers but tended to be discounted within zymotic theory. As Coleman argues, to assert the priority of the atmosphere as the cause that could incorporate all the facts about diphtheria served to "maximize the influence of factors beyond the control and probably beyond the comprehension of ordinary persons."[30] He further argues that the way was thus opened for the specialist (i.e., the sanitarian) who could analyze these environmental factors. Yet the focus on the atmosphere that some sanitarians favored "lent itself poorly to exact study." It left much unexplained. European researchers continued to look for the cause of diphtheria among the plethora of microorganisms in the throat, while physicians in New York City began to look more closely at other possible sources of the cause of diphtheria. In the ensuing years, New York City physicians gave greater attention to the treatment of the disease.

Treatment and Prevention

By the 1880s, physicians in New York City defined diphtheria as an amalgam of both Bretonneau's views and the zymotic theory. It was a specific disease, associated with the production of a pseudomembrane caused by a parasite, whose growth and multiplication was favored by filth. Confusion with other throat afflictions was still common, as were debates about its etiology and transmission.

The treatment of diphtheria received the greatest attention by New York City physicians in this period. Announcements of cures appeared regularly in medical journals; however, the efficacy of these touted remedies was questionable. For parents of children with diphtheria the situation was troubling because of the variety of remedies that physicians applied and because of the violent character of some of those remedies. In general, treatment focused on the patient's environment, the pseudomembrane, and, finally, the effects of the disease throughout the body.

When a case of diphtheria was diagnosed, physicians recommended that the sick child be isolated at once if possible. Physicians employed two kinds of treatment to deal with the local and constitutional symptoms of the disease. First, they attempted to dilute, wash away, or forcibly remove the pseudomembrane. Sprays of limewater and other alkalines, such as pepsin, typsin, and papayotin, were used to try to dissolve the false membrane. In desperation, some inexperienced physicians even attempted to forcibly remove the membrane, though such a procedure was viewed by many as "a procedure which is to be mentioned only to be condemned."[31] Physicians typically tried to use some remedy or method that was capable of dissolving the false membrane without irritating or causing injury to the surrounding tissues. Antiseptics and bactericidal agents were employed in order to destroy the poison produced by the infection. Chloride of iron was claimed to be effective because of its local astringent and antiseptic properties, as well as for its effect on the systemic symptoms produced by the poison.

Mercurials were widely used for the constitutional symptoms of the disease, and their effects were just as widely debated. Calomel and bichloride of mercury were given internally. Hypodermic injections of mercury were employed in some cases as well. The more experienced physicians urged caution in the use of mercurials because of their known debilitating and poisonous effects on the body. Such caution was not always heeded, however. As prominent New York physician Dr. Joseph Winters noted, one physician "administered calomel in forty-grain doses in serious cases of diphtheria, and one hundred grains in twenty-four hours. It almost constitutes malpractice to give such a dose to a young child depressed by the poison of diphtheria."[32]

Winters was clinical professor of diseases of children at the University of the City of New York and author of *Diphtheria and Its Management* (1885). Winters's book along with Abraham Jacobi's *A Treatise on Diphtheria* (1880) and C. E. Billington's *Diphtheria: Its Nature and Treatment* (1889), were three of the most widely cited texts on diphtheria in New York City throughout the 1880s and 1890s. The three generally concurred that physicians' treatment of diphtheria had become too aggressive. Both Dr. Winters and Jacobi argued that the use of small doses of mercurials like calomel was quite sufficient for most cases of diphtheria. The real problem, as Winters noted, was that their use had become indiscriminate. There was little evidence, in his opinion, that they had any specific effect on the course of a case of diphtheria:

The too universal extolment of them at the present time has caused an indiscriminate use of them, which brings harm to our patients, and will in the end cause the abandonment of a drug which is of undoubted utility if used as an auxiliary to our treatment, and will bring just opprobrium on our profession for using a remedy which has been over-zealously recommended on no sufficient foundation, and without any criterion to guide us.[33]

Winters's comments point to the role that therapeutics had in maintaining the professional authority of physicians in an era when orthodox medicine was routinely challenged by alternative practitioners. If treatments were used indiscriminately, without adhering to some guiding principles, then physicians' credibility would certainly be undermined. Many treatments appeared to be effective when used in the early stages of a case of diphtheria, and all observers reported instances when their applications of antiseptics and astringents had been successful. Yet in severe cases there was no one treatment that physicians could rely on.

Winters was fully aware of the physicians' failures with diphtheria patients as well as their need to appear competent. Some cases would do well under almost any form of treatment, while other cases seemed not to be affected at all. In these cases, physicians were prone to tell parents that the case was one of "unusual virulency" or "malignancy." Rather than continue with this subterfuge, he proposed "a system of rational therapeutics to meet the indications of each individual case, based on well-considered physiological and pathological data."[34] Each method of treatment should be evaluated on the basis of indications for cure as measured by the age of the patient and the appearance of other cases; the presence, character, and location of the pseudomembrane; the range and severity of the complications; and the season of the year. According to Winters, every case had to be evaluated individually according to these characteristics.

He viewed with skepticism treatment that aimed to attack the poison or germ that caused diphtheria:

The mode of treatment in vogue has for its object (according to its exponents) to attack the poison or germ of the disease, as an entity, and destroy it, leaving out of view every other consideration. Physicians in search for such a specific are distracted from a rational treatment. These experiments have doubtless killed more patients than germs.[35]

Winters argued that a judicious combination and sequence of treatment, "adapted with judgment to particular cases," was the only rational approach to the treatment of diphtheria.[36] His approach was hardly a new one. American physicians' therapeutic perspective had long stressed the need to match treatment to the "idiosyncratic characteristics of individual patients and their environments."[37] The push for a rational therapeutics at this time merely underscored the continuing failures of the treatment of diphtheria as its incidence increased. Winters's skepticism toward bacteriology reflected his concern that diphtheria was beginning to be seen as a disease defined by the germ that produced it rather than by its specific symptoms and effects throughout the body.

While mild cases of diphtheria responded to a variety of treatments, severe cases were handled either by tracheotomy or intubation. Surgical intervention in severe cases of diphtheria was usually done when the patient's breathing was blocked and asphyxia appeared to be imminent. Until 1887, physicians usually attempted a tracheotomy to save these patients. The procedure involved making an incision in the trachea and inserting a tube in order to open an air passage that had been closed by the pseudomembrane. Physicians had varying rates of success with the procedure. Jacobi did not put much faith in it. Between 1860 and 1887, though he performed more than seven hundred tracheotomies, the survival rate of his patients was very low.[38]

In 1881 New York physician Dr. Joseph O'Dwyer introduced a new method for keeping the throat open in severe cases of diphtheria. He inserted a small hollow tube, later known as the O'Dwyer tube, into the throats of his patients. O'Dwyer's tube was widely viewed by New York physicians as a better and more successful method than tracheotomy. Jacobi claimed that after 1887, he "rarely ever operated, and my friends stopped tracheotomy when O'Dwyer taught us all intubation."[39] O'Dwyer's method of intubation was widely recognized as one of the most effective surgical interventions in the treatment of diphtheria.

Individual Prophylaxis

In the United States, although little laboratory research on diphtheria was being conducted before 1890, physicians prided themselves on what they considered to be their achievements in diphtheria prophylaxis. In his *On the Treatment of Diphtheria in America*, Jacobi claimed that "prophylaxis appears to be more anxiously attended to in the United States than elsewhere. All over Europe the textbooks or monographs contain no or but scanty allusions to prevention."[40] For physicians, prophylaxis focused on the individual. New York physician Augustus Caillé supported Jacobi's view. He argued that "if the nasal and oral cavities are kept tolerably clean by means of harmless, non-irritating liquids known to possess antiseptic (disinfectant) properties, the frequency of diphtheritic infection is markedly reduced."[41] Caillé recognized, however, that his emphasis on prophylaxis might not be taken as seriously as it deserved to be by physicians.

The conception of diphtheria as a zymotic disease implied that the physician had limited capabilities in prevention of the disease and ultimately had to turn to public health authorities. While sanitary science had its place, Caillé argued, the utmost attention must be attached to personal and individual preventive measures. In Caillé's view, physicians had little choice but to pursue preventive measures since there was still much that was not known about diphtheria despite the increasing search for a bacterial cause. He wrote:

> If we admit our ignorance of the true nature and significance of so-called diphtheria, we must also admit that we cannot with certainty differentiate clinically between contagious and non-contagious acute inflammatory changes of the mucous membrane. . . . Furthermore, we may safely commit ourselves to state that, although the line of treatment which we pursue in diphtheria is not without its influence in checking the progress of the disease and stimulating the system until the disease shall have exhausted its virulent properties, we have no means of absolute control at our command, or in other words, no specific treatment for diphtheria.[42]

While the zymotic conception of diphtheria implied that no individual could avoid the risk from diphtheria, Caillé argued that individual prophylaxis could certainly reduce the risk. Without the absolute control that an effective treatment of diphtheria potentially held, he wanted physicians to at least attempt to control the individual practices of their patients. This focus on individual prophylaxis had increasing implications for the control of diphtheria by public health authorities. These implications were articulated most forcefully by Abraham Jacobi.

The Death of Ernst Jacobi

Diphtheria had both professional and personal implications for physicians. Their experiences watching the deaths of young children from diphtheria left lasting impressions. Victor C. Vaughan, a physician who would later become the first professor of hygiene at the University of Michigan, counted one of his experiences in an outbreak of diphtheria among the "most distressing pictures covering the walls of the memory chamber of my brain by the invisible hand of epidemic disease."[43] In 1886, he had seen diphtheria "with every possible complication and sequel" in a village of some two hundred families located in a pine forest in Michigan:

> The physician of the community had never seen the inside of a medical school and when a child died the body was taken to the church, the school children filed in and each kissed the corpse. This doctor had reported to the State Board of Health the prevalence of a new noncontagious disease. The memory of my experience in this village is one of the distressing pictures to which I have referred (emphasis in original).[44]

While Vaughan's anger was directed at the failings of a particular physician, in some cases physicians turned that anger upon themselves. Even the best-trained physicians often watched helplessly as their own children succumbed to diphtheria.

On Sunday, June 10, 1883, seven-year-old Ernst Jacobi died of diphtheria.[45] This death was notable because Ernst was the son of two of New York's most eminent physicians. His mother, Mary Putnam Jacobi, was the first woman graduate of the École de Medicine (1871), a winner of the Boylston Prize, and the first female physician admitted to the New York Academy of Medicine. His father, Abraham Jacobi, then clinical professor of the diseases of children in the College of Physicians and Surgeons, usually referred to as the father of modern pediatrics in the United States, was one of the foremost writers on diphtheria in the country.[46] Yet even their knowledge was not sufficient to protect their son. Mary Putnam Jacobi wrote in 1883 to a friend of her pain over the loss:

> But I never should feel that any knowledge now gained or even any lives saved, would be in any way a compensation for Ernst. I have rather a secret morbid longing to hear of everyone losing a first born son as they did in Egypt and had the death list every day to pick out the cases of seven and eight years. I feel sometimes as if the whole world must stand and [look?] at me for my inability to [save?] this lovely child. This [   ] feeling of self contempt perhaps in one way helps me to bear this terrible loss.[47]

Abraham Jacobi was equally distraught. Ernst was the first of his children. to live to the age of seven. One biographer noted that he never recovered from the death of his son.[48] Ernst's death raised a troubling question for the Jacobis. How could their child have contracted diphtheria? Certainly they had long experience in diagnosing the disease. Abraham Jacobi had seen hundreds of cases of diphtheria in the years he had practiced in New York City. The terrible toll that the disease had taken on the children of the city had proven to him the value of prophylactic measures. "We have seen plenty of sadness," he wrote, "but also many opportunities to learn."[49] The Jacobis routinely practiced the measures that Abraham Jacobi had long recommended for the prevention of diphtheria. His dictum was "A healthy child with healthy mucous membrane is not apt to be taken with diphtheria." To him, prophylactic measures were not simple palliatives; indeed, they had "proved a great safeguard, and have protected [us] from still greater calamities than we have endured."[50] Yet even these methods, rigorously applied, had not saved his son. While Mary Putnam Jacobi considered intensifying her studies of the etiology of diphtheria in order to make sense of Ernst's death, Abraham Jacobi turned to questions of how diphtheria was transmitted.

"The Scourge of the Community": Adults with Mild Sore Throats

Jacobi's 1884 article, "Diphtheria Spread by Adults," written just a year after Ernst's death, is a polemic about the cause and transmission of diphtheria.[51] In building his case for the idea that diphtheria could be spread by adults, he challenged current views held about diphtheria among the community of elite physicians in New York and those in the larger medical community.

First, he argued that diphtheria did not arise spontaneously from moist walls of houses or the dry dust of streets as some had claimed. While most physicians spent much time looking for the predisposing causes of diphtheria in sewage, the traps of water closets, and plumbing in the cellars of their patients' houses, Jacobi felt they were avoiding the more important, direct sources of contagion of diphtheria: the noses and throats of their patients. "There is but one predisposing element, viz., *a sore mucous membrane*," he wrote, "and but one cause of an individual attack of diphtheria, *viz. direct contagion*."[52]

In his earlier *Treatise on Diphtheria*, published in 1880, Jacobi had discussed the definition of diphtheria, and by 1884 he had in his own mind settled the issue. He affirmed the view that the definition of diphtheria was not in question as long as the characteristic membrane was present. In his view, the definition of diphtheria did not depend on the size or location of the membrane, nor was it limited by the character of its symptoms. By this he meant that a patient could have all of the symptoms associated with diphtheria—increased temperature, difficulty swallowing, and glandular swelling—in either mild or severe forms. As long as the characteristic membrane was present in some form, it was a case of diphtheria.[53] Jacobi stressed that the disease was always dangerous for the person who had it, but that it was also much more dangerous by its "potential influence," its contagiousness:

> No matter how slight a case may be at a given time, it can extend to the nares or the larynx; or terminate in paralysis or death. Besides all that, it is communicable to another person; a mild case may as well generate many severe ones as a severe one need not necessarily be the source of misfortune to others. For neither in diphtheria nor in other contagious diseases is contagiousness the unavoidable sine qua non.[54]

Diphtheria, Jacobi asserted, appeared in mild and severe forms. He emphasized that this disease, like other contagious diseases, did not necessarily strike every person in its path: "While a cannon-ball, however, may miss, a pistol bullet may carry destruction."[55]

In this case the "pistol bullet" he had in mind was mild cases of diphtheria. He referred specifically to cases in the earliest stages of the disease where the pseudomembrane was not well formed. He had observed many such cases in adults. In these cases physicians tended not to make a diagnosis of diphtheria but usually called such cases follicular tonsillitis or some other form of sore throat. Jacobi strenuously objected to this classification. These cases, he argued, should be called, and were by his experience, diphtheria:

> What to-day looks like a point, or four or five points covering the outlets of ducts, may tomorrow be a confluent membrane. Just as well, you might withhold the name of variola from a case of smallpox so long as it was not confluent or did not destroy life. You have a family sick with affections of the throat and nose; a child is dying of laryngeal croup, another of nasal diphtheria, glandular swelling, and sepsis; others have severe pharyngeal affections; others but slight tonsillar tufts, which may or may not coalesce and extend into membrane. Is the first one "croup," are the others "diphtheria," is the last one "follicular amygdalitis"? Or the first case in a family was just such

a one as was denominated tonsillar folliculitis by the family physician, and no pre-
ventive measures were taken to protect the rest of the little flock. From that moment
it was that the extermination began in that family. We have seen it all; we ought to
heed it all.[56]

Jacobi had seen many cases of diphtheria, particularly in adults, where the symptoms consisted only of punctate deposits (marked points or dots) on the membranes of the tonsils, and where the patient had a little muscular pain and some difficulty swallowing. Usually the symptoms were not severe enough to prevent the person from performing a normal routine. Yet, he argued, despite the mildness of the symptoms, these adults did have diphtheria and did transmit it to others. The case he offered as an example was one very familiar to him and is worth quoting in its entirety:

In the family of a physician there were two children — a boy, in good health otherwise;
a girl, robust and vigorous. Both suffered from diphtheria repeatedly for many years,
until the girl came near dying, and the boy died. What was the cause of the constant
attacks dragging over years? The fact had been overlooked that it so happened that
almost every time when the children were sick the old, trusted, and trustworthy nurse
was also affected with diphtheria, sometimes seriously. What, however, was not
known at that time, and was discovered later, indeed too late, was that the woman
had concealed many an attack of throat disease, fever, difficult deglutition, out of a
sense of duty; that she had often repeated those medicines which she had been sup-
plied with before. In the early part of the summer of 1883 the boy died; the surviving
girl was sent to the country; also the nurse who had been sick with diphtheria for
weeks. While in the country the child was not in so close and intimate company with
her nurse as in the city and in winter. They returned to the city in September. In Octo-
ber, the child was taken sick with diphtheria, the nurse having taken "throat medi-
cine," unknown to anybody, for some time. Then the nurse was discharged; that was
the end of the boy, and of diphtheria in that family.[57]

The similarity of the facts in this case and the experiences of the Jacobis in the death of their son Ernst is striking.[58] After Ernst's death, the Jacobis had sent their family nurse back to Germany. They believed that her frequent sore throats, which she had concealed from them, were the source of diphtheria in their household. Jacobi argued to his colleagues that "those . . . who feel well, or well enough to be about, are the scourge of the community at large."[59] Not only nurses, but all domestic help were dangerous in his view:

"Thus, if there be any class of persons who are the constant transmitters of diphtheria, and require attention and caution, it's nurses and cooks, teachers, hair-dressers and barbers, shopkeepers, restaurant keepers, and all those people who are in constant contact with all classes and ages."

Oddly enough, Jacobi did not see physicians as the same order of threat as the groups mentioned. Obviously, physicians too came into contact with "all classes and ages" especially in a city like New York. Yet Jacobi felt that the physician enjoyed "a certain degree of both active and passive immunity," based on the number and duration of his exposures to diphtheria. This relative immunity, he argued, defined whether the physician, too, could become "more or less both endangered and dangerous." The assumption he made was perhaps that physicians would not be so careless in their behavior, once they had contracted a sore throat, as those in his designated dangerous classes, though there were many instances cited in the literature of cases of diphtheria among physicians and nurses. Jacobi concluded this paper by lamenting that the insights he had now gained had come to him too late, especially too late to save his own child: "It has dawned on me too late that it is just so with diphtheria, and that preventive measures will be more effective when we look for the cause of every case of diphtheria in the nares and throats of living persons. There is as much diphtheria out of bed as in bed; nearly as much out of doors as in doors."[60]

Jacobi's article on the spread of diphtheria by adults had introduced a number of new and radical elements in the discourse about diphtheria. First, he was insisting that diphtheria not be defined solely on the basis of a well-formed pseudomembrane. Second, he claimed that adults with mild cases of the disease could transmit it to others. In one comment on the paper, which appeared in the *Boston Medical and Surgical Journal* a month later, a physician from Plattsburg, New York, disagreed with Jacobi on a number of points. Jacobi's article, in his opinion, was full of "vagaries and contrary to clinical facts."[61] Dr. D. S. Kellogg argued that since cases of diphtheria often appeared in the country where no possible source of contagion could be found, this evidence suggested that the disease could arise spontaneously. He also did not accept the idea that follicular tonsillitis was diphtheria. He argued that there were cases of sore throats so mild and "peculiar" that it was difficult to tell what they were and that there were far fewer such cases than cases of diphtheria. Dr. Kellogg was most at odds with Jacobi over his comments on the class of people who should routinely be examined for diphtheria:

But why stop here? Why not station special examiners on all street corners, in all railroad stations, at all sanitary crossroads, and in numerous other places, so that those, "who are in constant contact with all classes and ages," and especially those, "who are out of bed and out of doors" may be kept from a possibility of spreading any possibly contagious disease? Perhaps some labor-saving machine could be invented and used that would facilitate these examinations.[62]

Dr. Kellogg was not alone in his opposition to Jacobi's views. There is no record that Jacobi responded to Dr. Kellogg, but he did answer questions posed by others. Two years after Jacobi's article was published, he addressed the questions raised in this earlier paper from his new position as president of the New York Academy of Medicine. The paper, entitled "Follicular Amygdalitis," was read before the Section of Theory and Practice of Medicine of the Academy.[63] Jacobi opened his discussion by noting that the German physician, Professor B. Fraenkel, had discussed his views before the Berlin Medical Society.[64] Fraenkel is quoted as suggesting that Jacobi's views went too far. He noted that he had seen a case where a twelve-year-old girl suffering from "angina lacunaris" had infected her whole family with diphtheria. Yet he felt such cases were rare. He conceded, however, that diphtheria could be spread in this way and that therefore such cases should be isolated as a cautionary measure.[65]

In his 1886 paper, Jacobi went to great lengths to clear up the questions and misconceptions about his position on the role of mild cases of sore throat in adults as sources of diphtheria. First, he noted that his earlier point was not that adult cases of sore throats *caused* diphtheria, but rather that physicians often misdiagnosed mild cases of sore throats that he felt should be called diphtheria. The misnaming, he noted, was "but a subterfuge for the lack of a correct or complete diagnosis." The diagnosis of diphtheria, as Jacobi well knew, continued to be a difficult problem in 1886. A certain diagnosis could only be made in severe cases where a clearly defined membrane was present. Since a diagnosis of diphtheria was never considered to be unimpeachable except in cases with a well-developed membrane, Jacobi argued that there was a large second class of cases that had been called by a variety of names. When an epidemic of diphtheria was prevailing, there were many such cases. Citing a number of discussions in the voluminous literature on throat afflictions, he still maintained that many of these cases were mild forms of diphtheria. He could not understand why his argument had been received with such skepticism. "Why diphtheria should be observed in the worst form only, when cholera, yellow fever, variola, and scarlatina are permitted to run a mild course oftener than merely occasionally," he wrote, "I cannot understand." He had observed this mild variety of diphtheria mostly in adolescents and adults. Repeating his earlier warning, Jacobi again argued that there was no escape from this "mild, murderous variety."[66]

It is clear from the discussion that followed Jacobi's paper that there was little consensus on the issues he raised. Dr. T. E. Satterhwaite argued, based on the research he had conducted for the New York City Board of Health in 1874, that diphtheria was often fatal, whereas follicular amygdalitis or tonsillitis was neither fatal nor contagious. Dr. C. E. Billington conceded that the onset of diphtheria was indistinguishable from follicular tonsillitis, yet that its subsequent course was quite different. Given this situation, even the experienced physician, he maintained, must reserve his diagnosis for a short time. Most of those commenting made reference to severe cases of diphtheria. Jacobi's views were not dismissed out of hand, but the physicians gathered tended to agree that in the doubtful cases to which Jacobi referred, it was simply impossible to make a diagnosis. They also objected to Jacobi's comments about the appearance of diphtheria in adult cases. His emphasis on the form of the deposits on the tonsils and throat was not seen as definitive. They did agree with Jacobi, however, that with these mild cases it was "better to regard them as belonging to the more dangerous form of the disease."[67]

These physicians recognized that their inability to accurately diagnose a case of diphtheria in its earliest stages often created a credibility problem for them with their patients. Dr. J. Lewis Smith noted:

> A supposed error of diagnosis is often made by physicians, always to their discredit, who diagnosticate catarrhal laryngitis, but who find after two or three days that their patients have diphtheritic croup (diphtheria). A considerable number of such instances have come to my notice, always with the ill-will of families toward their physicians.[68]

Abraham Jacobi had pointed out that despite the risk of alienating patients by changing a diagnosis, physicians had to carefully watch any case of mild sore throat in adults, particularly when an outbreak of diphtheria prevailed. The implications of Jacobi's insistence that mild cases were dangerous and that those suspected of hiding their condition should be routinely examined received a mixed reception.

Jacobi's reaction to his son's death compelled him to locate the origin of diphtheria distant from himself, his physician wife, and physicians in general. This gave him assurance that the true source of diphtheria lay in some external source of pollution. He believed the external agent lay in filth and in persons who lived in the tenements of New York City. In Jacobi's view, the existence of these dangerous classes, the carriers of contagion, exposed the limits of individual prophylaxis and the boundary between the endangered better classes and the dangerous immigrant classes. He saw no alternative except a massive program of surveillance. He envisioned city authorities conducting the routine examination of the throats of all children and, in times of epidemics, the examination of people in every public place in the city along with the disinfection of all public transportation. Would such practices be considered a violation of the rights of citizens? Yes, he answered, but in such cases, "It's not the society that tyrannizes the individual; it is the individual that endangers society."[69]

Jacobi could not marshal support for such a program of surveillance from the ranks of his physician colleagues. How such a program for the identification and control of mild cases of diphtheria could be conducted was a question he could not answer. The continuing problems in the diagnosis of mild cases of diphtheria and the fact that the recognition of mild cases was highly subjective and variable provided a thin foundation on which to base such practices.

Public Health

Physicians engaged in public health work were particularly cognizant of the prevailing sense of pessimism regarding the possibility of controlling diphtheria. At the annual meeting of the American Public Health Association in 1890, Dr. G. C. Ashmun presented the report of the committee appointed by the association to examine the cause and prevention of diphtheria. They noted that the public, including medical practitioners, had grown "apathetic and hopeless in regard to the treatment of diphtheria to a degree not experienced toward any other disease."[70] This attitude was due in part, they suggested, to the wide distribution of cases and the constant presence of the disease throughout communities. The apathy was compounded by the lack of any satisfactory preventive measures that could be employed effectively.

Public health physicians were not content to wait until medical science could provide them with some specific remedy to control diphtheria. They saw the failure to mount an energetic program against diphtheria as a "constant reproach to sanitary science." The committee therefore polled physicians in eight states on questions about the cause, the treatment, the mode of transmission, and the effectiveness of isolation and quarantine. Several of the responses to these questions are illustrative.

It was generally accepted by health officers that diphtheria was caused by a specific agent or germ. However, a majority of those polled believed that diphtheria was *not* caused by "germs, ptomaines, or products and conditions developed within the body." Considering germs as a source of diphtheria could not account for the presence of the disease in isolated cases where no source of contagion could be identified. Air, water, and food were identified as the media by which the diphtheria virus entered the body.

There was a consensus that the research on the etiology of diphtheria had generated such a body of conflicting facts that "a very conservative and even doubting state of mind" existed among many practitioners. However, the committee felt that some very important and useful information about diphtheria had been discovered by laboratory research. These public health experts' proposals for the control of diphtheria were commensurate with those of physicians who dealt with individual patients and with zymotic theory in general. Their recommendations were in complete agreement with those of the New York City Health Board's existing methods for controlling diphtheria.

Public health experts believed they could control the disease if they aggressively pursued cases and rigorously applied disinfection or, failing that, if they could shake the public out of its "submissive inaction." However, they failed to consider the impact of declaring all cases as "dangerous" sources of contagion. Physicians would be leery about reporting cases of diphtheria in a context where the results of such reporting invoked such aggressive measures. The committee had recommended that cases be quarantined for as long as four weeks after all signs of infection had disappeared. Given that everyone recognized the difficulty in diagnosing diphtheria, physicians could easily claim that they could not identify a case with certainty, especially if they wanted to protect the patient and family from the intervention of municipal authorities.

These suggested public health practices for the control of diphtheria would receive much comment and refinement during the last decade of the nineteenth century. But a significant impediment faced by public health experts attempting to implement the committee's proposals was the wide prevalence of diphtheria, particularly in urban areas. Physicians expressed some confidence that patients from the "better" classes could be educated to protect themselves from diphtheria by the disciplined application of appropriate prophylactic measures. They held to this position even though there were striking and inexplicable instances where it had failed. Public health experts made a distinction between what was appropriate prophylaxis for the better classes and the needs and perceived behaviors of the growing poorer, largely immigrant population in the cities.

Epidemiology of Diphtheria in New York City

The relatively small physical size of Manhattan Island (40.34 square miles) has always belied its influential position in American life. In the late nineteenth century it was the wealthiest, most populous, and busiest metropolis in the United States. It was the leading center of business and cultural life in the country. Its port handled nearly 50 percent of the nation's foreign trade. In the next ten years, it would possess the world's largest telephone system. Every form of transportation could be seen in its crowded streets. Steam-driven elevated trains traveled over streets clogged with cable cars, more and more automobiles, and countless bicycles. Garbage often made travel more difficult as the notoriously corrupt city street-cleaning department barely managed to keep the streets cleared.[71] It is perhaps a cliché to say that New York in 1890 was a city of extremes. Physically, the lower part of the island contained districts whose population density exceeded comparable sections in Bombay or Prague.[72] The Upper West Side, on the other hand, was indistinguishable from the outlying countryside. The cosmopolitan character of the city was reflected in its population of 1,515,301, which included approximately a third native-born whites, a sizable number of African Americans; more Irish than any city in Ireland; Scotch and French; Germans, Russians, and Poles; Canadians, Scandinavians, Hungarians, Bohemians, and Italians; and smaller numbers of Chinese, Syrians, and Arabs.[73] In sum, immigrants and their children made up four-fifths of the population of New York City in 1890.[74]

As the population of immigrants continued to increase, adequate housing became a critical concern. Observers of city life, especially native-born elites involved in charity organizations, grew increasingly concerned about the glaring disparity between the "promise of American life and the reality of the tenement house slum."[75] Tenement housing was overwhelmingly occupied by immigrants. In 1890 nearly 67 percent of the city's population lived in housing containing twenty-one or more persons, and more than 160,000 of the tenement dwellers were children under five.[76] New York had the largest population living in tenement housing of any city in the northeast United States.[77]

For many of New York's elites, the most difficult aspects of the increasingly complex urban environment of New York were embodied in the twin evils of immigration and the tenement house slums. The tenement was for them "the symbol of pathological social conditions as well as a source of sickness and death."[78] To elites the three D's — Dirt, Discomfort and Disease — were the striking features of this environment.[79] To these observers, dirt and overcrowding equaled disease. With an environmentally based sanitary science as well as their own observations to guide them, they were not surprised to see high rates of all diseases in the tenement house districts. Yet an analysis of death rates across the city, especially with regard to diphtheria, countered common assumptions.

The most comprehensive analysis of the death rates in the city due to various diseases is found in the special report appended to the 1890 census authored by Dr. John S. Billings Sr., then a special agent to the U.S. Surgeon General's office. In 1890, diphtheria was the sixth leading cause of death in the city.[80] Over the six years covered in the report, there were 15,199 deaths due to the disease, giving an overall rate of 181.63 per 100,000 population. In the first four months of 1890 alone, there were 1,870 deaths due to diphtheria and croup, representing a rate of 123.41 per 100,000. Since diphtheria primarily affected children, it is more illustrative to note the age distribution of deaths. The death rate for children under five was nearly nine times greater than for those between ages six and fifteen years.

The highest death rates occurred among Italian and Bohemian immigrant children. The highest death rate due to diphtheria during the census period was 302.88 per 100,000 in District C of Ward 19. A largely Italian district with some Russian and Polish blocks, Ward 19 was located on the East Side of the city, bounded by East 50th Street, the East River, East 40th Street, and Second Avenue. The second-highest death rates occurred in District I of Ward 19, largely populated by Bohemian immigrants and located about twenty blocks north of District C. The diphtheria death rates were high in every district along the banks of the East River from 14th Street to 120th

Street. Physicians working in dispensaries on the East Side referred to these tenements as "diphtheria nests" because of the consistently high number of cases that occurred no matter who lived in these buildings.[81]

Yet overcrowding alone could not explain the high rates. The densest wards in the city were the Tenth, Thirteenth, and Eleventh, the so-called Jewish wards on the Lower East Side. The diphtheria death rates in these wards, however, were only slightly higher than the city's average. The heavily Italian Fourteenth Ward shared a border with the Jewish Tenth Ward, yet it had higher mortality rates.[82] This unexplained difference in rates is particularly evident in comparisons of children under five years. Jewish wards had higher numbers of children under five years of age, those most vulnerable to diphtheria, yet the death rates due to the disease were lower than those of their neighbors, especially the Italians and Bohemians. All three groups had diphtheria death rates higher than the average for the city. In sum, the Lower East Side districts were extremely crowded in close proximity to each other, yet diphtheria death rates differed inexplicably.

This disparity in mortality rates is attributed to a number of causes by both contemporary and modern writers.[83] Contemporary observers believed that the generally lower death rates in Jewish wards were due to the rarity of alcoholism and the influence of religious law and social customs of the Jewish people. One health officer noted, "The rules of life which orthodox Hebrews so unflinchingly obey, as laid down in the Mosaic Law . . . are designed to maintain health. These rules are applied to the daily life of the individuals as no other sanitary laws can be."[84] The Italians, by comparison, were routinely criticized for their lack of cleanliness and for living in close quarters, with two to four rooms per family and sometimes two families per apartment. The high rate of illness among Italian children was attributed to the fact that they were born and raised in the tenements, were poorly nourished, and grew to be anemic.[85] It was also argued that the Italian immigrants had been physically affected by the difference in the climate between their native Italy and New York City. Observers attributed their relatively greater susceptibility to disease to the change in climate and the characteristically low protein diet of this community.[86] Contemporary observers made little attempt to establish more definitive reasons for the differential death rates to diphtheria among immigrant groups in the city. While physicians and sanitarians knew that the relationship between unsanitary conditions and diphtheria was a weak one, few believed that the deplorable conditions of the tenements and the overcrowding were not a contributing factor in the prevalence of the disease. Jewish dietary habits were seen to fit well under the rubric of personal cleanliness that was advocated as a preventive measure against diphtheria, whereas the perceived lax habits of the Italians did not.

Despite the perceptions of contemporary observers, modern writers have argued that the relationship between overcrowding, high mortality, age structure, and the ethnic composition of immigrant neighborhoods in late-nineteenth-century New York was a complex one. There were no firm boundaries between these neighborhoods, as the contemporary statistical data implied. No simple comparison between any of the variables—age, ethnicity, or density of population—can adequately explain mortality differences across New York City at this time.[87]

The Impossibility of Control

In the thousands of tenement houses of New York, in each of which from ten to thirty families are huddled together on a building space of 250 square metres, every case of diphtheria becomes a danger to the story, the house, the neighborhood, the city. Isolation becomes an impossibility, a single outbreak means a heap of misery and a number of deaths.[88]

The impossibility of controlling diphtheria or any communicable disease in the conditions of the tenements was obvious to all observers. The control of diphtheria was predicated on the practices of personal hygiene, isolation of the sick, and ventilation and disinfection of the sickroom. In tenement housing, such practices were simply impossible. The overcrowded conditions of the tenements and the impossibility of providing separate rooms for the sick are made abundantly clear by the striking photographs of Jacob Riis, the preeminent chronicler of the conditions in the tenements.

One possible remedy to the situation was to move cases of contagious diseases that could not be properly isolated in the home to the hospital. Physicians in New York City had long been concerned about the need for contagious disease hospitals in the city, especially for the immigrant population; however, until 1885 there was no hospital in the city that would admit a patient with diphtheria.

In the spring of 1885 the health department opened a hospital at the foot of East 16th Street, named the Willard Parker Hospital in honor of Dr. Willard Parker, vice-president of the first Metropolitan Board of Health and a leader in the efforts to secure funds for the new hospital.[89] Located "sufficiently distant from inhabited dwellings to minimize the danger from contagion," it was devoted exclusively to the treatment of diphtheria and scarlet fever.[90] By 1890 the Willard Parker Hospital could accommodate seventy to eighty patients. The hospital received its funds from the city's health department and was under the immediate charge of a house physician who received his directions regarding treatment from a consulting board of physicians. This board consisted of several of the leading physicians in the city involved in the treatment of diphtheria: Abraham Jacobi, Dr. George F. Shrady (also the editor of the influential journal *Medical Record*), Dr. E. G. Janeway, Dr. W. Stimson, and Dr. Joseph O'Dwyer.

In lobbying for funds for the hospital, Jacobi and Shrady emphasized the escalating number of cases of diphtheria in the city: "The disease has existed as an epidemic in this city for the last thirty years. In 1884, there were 2,223 cases with a mortality of 1,090. . . . At the same ratio in 1888, 6,220 cases might be expected with a mortality of 3,000."[91] The physicians' appeal included Jacobi's warning of the danger posed to the city by many unrecognized and unacknowledged mild cases of diphtheria.

Jacobi and Shrady wanted New Yorkers to understand that diphtheria was not like smallpox; people could not be protected by exposure to the disease or by vaccination. Isolation of the sick was therefore critical to controlling the disease. Every effort must be made, Jacobi argued, to enable people to isolate the sick. Those who would not isolate themselves—for example, servants—should be prevented from going about their daily business because of the danger they posed to the public. In the presence of such a clear danger, Shrady argued, "We cannot vaccinate people to prevent diphtheria. We can only prevent it by taking them away from the sources of contagion and isolating them where they can be safely treated."[92]

While Jacobi and Shrady recognized that one hospital with seventy beds was insufficient to prevent the spread of diphtheria in a city where, in 1890, over four thousand cases had been reported,[93] they appealed to the public's concern for the children of the poor. And at least the principle of isolation had been validated by the establishment of one facility.

For the poor, the Willard Parker Hospital was a forbidding institution. The hospital's regulations were harsh. Mothers were allowed to accompany their children to the hospital only if they remained until the child recovered and provided that they complied with the rules and regulations of the hospital. They were required to assist the attendants of the hospital in duties connected with the care of patients. The health department insisted on strenuous methods to prevent the spread of contagion. Friends and other family members were allowed to visit the hospital only once each week and to remain for half an hour, provided that "they change their apparel and take such other steps as are deemed advisable to prevent the spread of contagion."[94]

As noted in the Annual Report, the health department was aware that there was a great deal of fear about hospitals among the immigrant population. Many parents refused to let their children be taken to the Willard Parker. Health department officials noted that Italian families were especially reluctant to let their children be taken away.[95] For some families the trip to the hospital was an arduous one. Little Italy, located near 110th Street, was quite far from the Willard Parker Hospital at East 16th Street. Deeply held cultural beliefs about illness and treatment of the sick; fear of dying away from friends and family; dread of hospitals with unfamiliar language, foods, and strange rules; and a lack of faith in public officials led many Italian families to conceal cases of diphtheria from health department inspectors. With respect to contagious diseases, one observer noted that Italians tended to respond cautiously. They avoided hospitals because they generally believed that diseases tended to run their own course helped along by traditional remedies.[96] A final source of ambivalence toward hospitals was the stigma attached to them as charity institutions.

There was an additional economic factor that led many families among all poor ethnic peoples to conceal cases of contagious disease: inspectors visiting families to investigate cases of contagious disease also reported on other violations of the city's sanitary code, including violations of laws regarding work in the home (i.e., the illegal sweatshops). A report of a case of contagious disease often led to all work being stopped in an apartment, with the attendant loss of income for the family: "Frequently the visit of the inspector is reported upon his entrance to the block, and the work is hidden until after the inspection has been made. In cases of contagious disease, the door is usually kept locked; the time consumed in opening it permits the hiding of the work."[97] These practices indicate that many immigrant families simply wanted to protect their children and their source of livelihood from what was seen as the unwarranted intervention of the city authorities. As a result, in 1890 only 488 cases of the 4,350 cases reported were treated at the Willard Parker Hospital.

The poor were not the only ones who refused to report cases of diphtheria to the authorities. Hotel owners commonly concealed cases because they too wished to avoid the stringent rules requiring isolation and disinfection of their premises when a guest contracted diphtheria. Since there was no hospital in 1890 for the middle- and upper-middle-class travelers to the city, there was no place to isolate such cases. Travelers were thus forced to leave the hotel or conceal the nature of the case and run the risk of infecting others.[98] Inadequate provisions for the isolation and thus prevention of diphtheria in all classes of citizens emphasized the difficulty of controlling the spread of diphtheria in the city.

Conclusion

In 1765 Dr. Francis Home wrote in a commentary on croup, "This disease appears, in general, to be a very dangerous one, and the more so as it is silent in its progress, and gives no visible alarm till death is near at hand. The first stage of this distemper often passes unobserved, and before we see it, beyond all remedy."[99] More than one hundred years later, Dr. Joseph Winters argued that Home's view still accurately described his experience with diphtheria in New York City.[100] By 1890 physicians in New York City concurred. Diphtheria was seen as an exceedingly formidable, dangerous, and whimsical disease. The public became increasingly aware of the specific danger it posed: the increasing incidence, the inexplicable epidemic outbreaks, and the failures of private and public medicine to protect its innocent child victims. Between 1850 and 1890 the awareness of the dangerous nature of the disease became conflated with the increasing visibility and perceived danger from the immigrant classes.

The process of defining diphtheria in the nineteenth century had begun with the identification of a characteristic set of symptoms coupled with a specific lesion, the pseudomembrane. Pathological anatomical research had shown that the pseudomembrane could not be the defining characteristic of diphtheria since it appeared in other diseases as well. Diphtheria appeared as a stable entity only in its most extreme forms, when the lesion and the characteristic symptoms were most visible. In its earliest stages, it continued to be named as some other throat inflammation, such as croup or follicular tonsillitis, even though the differences in clinical features and the pathological lesions were quite striking. Although in the majority of cases they conformed to a general type, the deviations from those types were in some instances so wide that many physicians believed they were seeing different or at the very least "mixed" infections.

All agreed that diphtheria was contagious, though there was no agreement on its mode of transmission. The diagnosis, treatment, and management of diphtheria was viewed as a complicated process by 1890. Physicians who wanted to save their patients and sanitarians who wanted to stop the spread of the disease found little solace in either the zymotic or clinical definition of diphtheria. They believed that diphtheria could be controlled if they could identify its specific cause, but they were limited to a conception of cause that could only be captured in clinically observable symptoms. While visible confirmation of diagnosis was possible in severe cases, mild cases presented symptoms shared by other diseases. Identification of persons with diphtheria was the cornerstone of an intervention that relied on isolation and disinfection. In the absence of definitive observable characteristics of the disease, their system of rational therapeutics, isolation, and disinfection failed to make the cause of diphtheria visible. Thus it continued its silent and dangerous progress.

2: THE PROMISE OF CONTROL

The discovery of disease-causing microorganisms radically transformed the identity of infectious diseases during the last decades of the nineteenth century. The introduction of the laboratory-based science of bacteriology is said to have solved the only outstanding problem with diphtheria, namely identifying which "parasite, germ, or ptomaine" caused the disease. But the issue was not only the cause but also its identity as a specific disease entity. This question was still at issue in 1890.

Though historians have noted that there was fairly widespread opposition to germ theory among American physicians, the response to the initial bacteriological research on diphtheria did not reflect a rejection of bacteriology itself. The initial hostile response reflected the failure of bacteriological research to answer those questions about diphtheria that were of greatest concern to physicians.[1] Their primary concern was to understand how to distinguish diphtheria from other diseases. Second, they sought to identify the source of contagion and its relationship to the disease. By 1890, when researchers in the United States took up bacteriological research on diphtheria, its indefinite conception raised problems for those at the bedside, at the dissecting table, and in the laboratory.

The laboratory was the direct instrument by which diphtheria was redefined with respect to microorganisms and bacteriology. Yet the laboratory is "never a mere instrument, it is also a practice which defines, limits and governs ways of thinking and seeing."[2] Diphtheria achieved a new, more stable identity through the laboratory. Yet this identity was fundamentally transformed in the process. The diphtheria that went into the laboratory was not the diphtheria that came out of it. As with other infectious diseases that were redefined with the discovery of microorganisms, diphtheria was now defined as being cause-based rather than symptom-based.[3]

This redefinition of diphtheria was at the center of debates over the status of bacteriological knowledge between regular physicians and those who were supporters of bacteriology. The status of laboratory-based knowledge in the United States was not firmly established by European research on diphtheria. This knowledge had to be confirmed under the specific conditions under which diphtheria prevailed in the United States. It was necessary to take the laboratory *to* diphtheria in order to take the disease through the laboratory. In New York City this was accomplished through the introduction of William H. Park's "culture kit." Park's kit and the program of bacteriological diagnosis that Hermann Biggs built around it took the laboratory to the streets of New York. The kit served to establish the diagnostic laboratory as a new authority in the control of diphtheria and other infectious diseases. The diagnostic program transformed the discourse of control in the history of diphtheria as well. Supporters of the new program articulated the control of diphtheria in more instrumental and reductionist terms than had the previous generation of experts. At the same time, the power of the laboratory was defined by its impact on their ability to control diphtheria, which they believed had the broadest possible implications for the control of all infectious diseases.[4]

Park's culture kit served as a *boundary object* between the laboratory, the physician, and the patient with diphtheria. Boundary objects are those scientific objects that both inhabit several intersecting social worlds and satisfy the divergent informational requirements of each of them.[5] Park's kit put bacteriological knowledge into the hands of physicians; it provided new information about diphtheria to the research and diagnostic laboratories; it allowed public health officers to more accurately document the conditions that promoted diphtheria in the city; and lastly, it created a new role for the state, first in diagnosing, and later in treating and preventing diphtheria. Thus Park's kit was key to the successful control of diphtheria in New York City.

The kit was also critical to the development of the first municipal diagnostic laboratory in New York City. If laboratory development in the nineteenth century can be thought of as a "chain of links that began with the laboratory devoted to basic research; followed by the clinical laboratory, which split its efforts between research and patient care; and ended with the ward laboratory, the workshop next to the patient, where knowledge and methods perfected in the other laboratories was applied,"[6] then Park's culture kit and the diagnostic laboratory that Biggs built up around it were crucial links in this chain and in the transformation of diphtheria into a disease defined by the laboratory.

This redefinition of diphtheria had significant concomitant implications for the relationship between physicians and the state. Because the diagnostic laboratory was under the auspices of municipal authorities, it threatened to usurp the physician's role as the sole diagnostician of disease, and it exposed the potential for the subordination of clinical diagnosis to bacteriological diagnosis.

The impact of the diagnostic program on the reception of the bacteriological conception of diphtheria and on relations between physicians and the New York City Health Department has long been overshadowed in historical accounts by the introduction of diphtheria antitoxin. Debates among physicians about the value of antitoxin must be understood in light of the redefinition of diphtheria that occurred as a result of the bacteriological diagnostic program instituted by Park and Biggs.

Bacteriological Research on Diphtheria

In November 1874 the New York City Board of Health requested that Dr. Edward Curtis and Dr. Thomas E. Satterhwaite investigate the causes and nature of diphtheria by micropathological examinations. Dr. Curtis was the honorary microscopist of the Board of Health, and Dr. Satterhwaite was a pathologist at Presbyterian Hospital.[7] Their work was one of the first experimental studies of diphtheria in the United States. They attempted to determine "the nature of the infectious principle of diphtheria, and . . . the circumstances that determine the infection." Since 1859 a number of researchers had reported on the presence of microorganisms in the human throat. In the case of diphtheria, such organisms had been seen near the false membrane. Satterhwaite and Curtis assumed that the bacteria found in the false membrane were the source of the disease. They focused their experiments on determining the presence or absence of bacteria in the membrane. They found bacteria in the false membranes they examined, but they could not distinguish between these bacteria and those found in a healthy mouth. Problems with their staining techniques and the limited power of their microscopes prevented them from making distinctions among the bacteria they observed.[8]

By 1880 many experiments were performed in European laboratories attempting to ascertain the relationship of bacteria to diphtheria. The German physician Edwin Klebs, who began his work on the study of microbes and disease during the Franco-Prussian war, turned to diphtheria in 1873. After a number of experiments, he identified a bacterium that he thought was the causal agent in diphtheria.

Friedrich Loeffler, Robert Koch's assistant, was the first to isolate the bacilli that had been identified by Klebs and to demonstrate their etiological role with respect to diphtheria.[9] Loeffler's paper, published in 1884, discussed the confusion that had surrounded research on diphtheria since Bretonneau. He attributed this confusion to "the nature of the disease, which was subject to great variations, which were due to the age of the patient, the virulence of the epidemics, the stage of the disease, and most of all to the complications which rendered it extremely difficult to investigate."[10] Loeffler accepted Bretonneau's views on the disease and focused his research on identifying the bacilli and establishing the nature of the "diphtheritic virus."

Using sections of the pseudomembranes of patients who had died of diphtheria and a special stain of methylene blue that he had developed, Loeffler was able to find the small rod-shaped microorganism described by Klebs. With a solid culture medium that included blood serum, he was able to grow abundant colonies of the bacilli.[11] Loeffler then inoculated guinea pigs, mice, rabbits and other animals with the bacilli obtained from the cultures. Although mice and rats were not affected by the inoculations, guinea pigs died with the characteristic pseudomembranes of diphtheria at the site of inoculation.

The etiological significance of the bacilli was evidenced in several aspects of the experiment. They were found in thirteen cases of clinical diphtheria. They lay in the oldest part of the false membranes and occupied a deeper position. The cultures were highly pathogenic for guinea pigs and small birds. The fact that the bacilli were found usually in small numbers only at the seat of inoculation but were never observed in the organs indicated that a poison was produced at the seat of inoculation that then circulated in the blood. Inoculation of the bacilli into the trachea of rabbits produced the typical false membrane seen in diphtheria.

Loeffler also discussed several findings that undermined the etiological significance of the bacilli. They were not found in all clinical cases examined. The typical arrangement of the bacilli in human false membranes was not found in experimental membranes. Attempts to infect animals by inoculation on uninjured surfaces were unsuccessful. And finally, the typical virulent bacilli were found in the mouths of healthy individuals. In sum, Loeffler argued that although the possibility that the bacilli he had isolated represented the long-sought virus of diphtheria could not be disproved, some doubt remained. He emphasized the need for further investigation, especially into the nature of the chemical poison the bacteria produced.

As subsequent observers have noted, Loeffler's work did not receive the same attention from bacteriologists and physicians as had Koch's identification of the tuberculosis bacillus in 1882. There are several reasons for this. Tuberculosis was surely the most serious disease of the nineteenth century; therefore, the discovery of the causal agent had garnered worldwide attention. Diphtheria, though a leading killer of children, had never held the same kind of public attention. But more important, Loeffler, unlike Koch, could not claim that the bacilli he identified were the undoubted etiological agent in the disease.

The criteria for etiological proof have come to be known as Koch's postulates. Though the criteria for the establishment of the relation of a microorganism to a disease were evolving during the period when Klebs and Loeffler were studying diphtheria, by 1884 Koch, through his work on the etiology of tuberculosis, had definitively established the set of criteria that had to be met. Briefly stated, they were:

1 *An alien structure must be exhibited in all cases of the disease.*
2 *The structure must be shown to be a living organism and must be distinguishable from all other microorganisms.*
3 *The distribution of microorganisms must correlate with and explain the disease phenomena.*
4 *The microorganisms must be cultivated outside the diseased animal and isolated from all disease products which could be causally significant.*
5 *The pure isolated microorganism must be inoculated into test animals and these animals must then display the same symptoms as the original diseased animal.*[12]

These postulates did not account for the nonpathogenic occurrence of bacteria.[13] On the basis of the postulates, Loeffler's conclusions that the bacteria he saw were the undoubted etiological agent in diphtheria were a bit guarded. The inconclusiveness of Loeffler's findings was significant in an era when the false starts and exaggerated claims by discoverers of pathogenic bacteria had produced a sense of skepticism among many physicians about the reliability of bacteriological research.

For physicians in New York, two aspects of Loeffler's work were problematic. Some questioned the evidence about the specific organism, while others viewed the finding of the bacilli in the throat of healthy individuals as proof that the causal agent had not been found. One of the earliest discussions on Loeffler's work among physicians in New York City occurred in the fall of 1884 at a meeting of the Clinical Society of the New York Post-Graduate Medical School and Hospital. Dr. Mary Putnam Jacobi opened the discussion with a report on Loeffler's findings. In the discussion that followed, most of the physicians present stated that they did not have the expertise to evaluate Loeffler's work. They could only comment on anomalies in the specific cases of diphtheria under their treatment. While not claiming to accept Loeffler's views, Mary Putnam Jacobi noted that no previous research had used the pure culture method and that therefore Loeffler had certainly come closer than anyone else in identifying the causal agent.[14]

A few months later, in the winter of 1885, in his inaugural address to the New York Academy of Medicine, Abraham Jacobi commented much more negatively on Loeffler's research. Jacobi saw Loeffler's work on diphtheria as simply the latest finding in a series of bacteriological experiments coming out of Germany that had not lived up to their claims of providing answers about the role of bacteria in disease. He felt that many American physicians had uncritically accepted many of these findings because of their inexperience in experimental work. Jacobi charged: "In America, also all of those who cannot judge of the question by their own investigations, that is, the practitioners, either general or special, have readily accepted the new gospel with but few exceptions." With respect to diphtheria research, he argued that the constant announcements that new bacilli had been found had made the entire project suspect:

Have we not had enough yet of the monthly installments of new bacilli which are invariably correct and positive sources of a disease, and replaced by the next man who comes along? Have we not yet enough of the statements, that, as for instance several bacilli are claimed to be the only cause of diphtheria, by several observers, that there may be several distinct bacilli every one of which can produce the same scourge? Is it not just as safe to still presume, that, when several forms of bacilli are believed to be such sole causes, that the real cause is neither?[15]

Regarding Loeffler in particular, he noted, "This time it is neither Klebs, or Eberth, but Loeffler. Reports and even editorials carry his name over the world. The very nature of diphtheria is said to be revealed again, as several times before; still, the discoverer admits there are cases without the bacterium. . . . The matter is becoming ludicrous."[16] Jacobi was much more convinced that pathological anatomical research offered the best hope of understanding disease processes. He believed that bacteric etiology too often begged the question as to whether organic or chemical poisons were the main causes of infectious diseases.[17] Jacobi was not alone in this view, which was particularly prominent among some German pathologists of the time.[18]

Bacteriological work on diphtheria continued in Germany and France, though Loeffler's work did not receive much direct attention for several years. Then, in the years between 1886 and 1890, several researchers confirmed his findings. In this next era, Americans would for the first time make significant contributions to the bacteriological study of diphtheria.

Bacteriological Study of Diphtheria Comes to New York

No systematic bacteriological work on diphtheria was being conducted in New York City or anywhere else in the United States before 1887. The physicians who followed the European work realized that the techniques involved in bacteriological research, particularly the use of solid culture media and the plate method of obtaining pure cultures required special equipment and training. Medical students interested in bacteriology typically arranged for postgraduate studies in European laboratories immediately following their formal medical school studies. Older physicians in the United States and those whose finances would not support a European trip were severely hampered in their attempts to gain more detailed information about the European work. There were few textbooks available in English, and the journal articles were printed in German or French. As McFarland notes, before 1884, when Klein's *Microorganisms and Disease* appeared, no book in the

English language treated the subject. There was also no standard apparatus available before 1882, and none made in America until about 1890. By
1890 only the merchants Eimer and Amend in New York had a catalogue
from which laboratory equipment used by Europeans could be easily purchased.[19]

Those physicians who had studied in Europe began, therefore, to fulfill a
variety of functions with respect to disseminating bacteriological knowledge
in the United States. Along with their new technical knowledge, they
brought back microscopes and other apparatus. Several of them translated
into English a number of important German and European texts in order to
teach the first courses in bacteriology at American medical schools.[20] More
importantly, a few key members of this elite group of American physicians
began, in this era, a transformation of American medical education and research based on the laboratory sciences.

After Koch's success in identifying the cause of tuberculosis in 1882, one
might assume that the interest among physicians in studying bacteriology
and building laboratories to continue such work would have been strong in
the United States. However, there was little national or local government
support for the building of such institutions. Only a few industrialists began
to devote some of their philanthropic efforts toward medical research. In
New York City in 1884, Andrew Carnegie donated $50,000 to improve the
small pathology research laboratory at the Bellevue Hospital Medical College headed by William Henry Welch.[21] Another small laboratory for pathological and bacteriological work headed by T. Mitchell Prudden was funded
by the Alumni Association of the College of Physicians and Surgeons in
1878.[22] In Brooklyn physician and industrialist C. N. Hoagland donated
$100,000 in 1888, after the death of a nephew from diphtheria, to establish a
laboratory dedicated to teaching and research in bacteriology.[23] The Marine
Hospital Service also operated a small laboratory on Staten Island in New
York Harbor. New York was therefore unique among U.S. cities by 1890 in
having three small but functioning laboratories for pathological and bacteriological research.

In 1885 T. Mitchell Prudden and William H. Welch became the first
Americans to study in the laboratory course in general bacteriology offered
by Robert Koch at the Gesundheitamt in Berlin. These two men, both grad-
uates of Yale and the College of Physicians and Surgeons, were destined to
become leading researchers and institution builders in medical research in
the United States. Prudden and Welch returned to America deeply influ-

enced by both the technique and explanatory power of the new science of bacteriology.[24] Welch returned to his new post in pathology at Johns Hopkins University.

Prudden, with much scantier resources, returned to his post as director of laboratory of the Alumni Association of the College of Physicians and Surgeons in New York. There he began conducting laboratory classes in the examination of tuberculosis tissue and sputum and the cultivation of bacteria on solid media. After his training in Germany in 1885, he revised the course in practical laboratory pathology to include the "study of bacteria in their relations to disease," the first course to be offered to medical students in the city.[25] Two years later the course was renamed Pathology and Bacteriology. By 1888, Prudden had successfully established the pathological laboratory at the College of Physicians and Surgeons, and he had also won a place for bacteriology in the medical curriculum. During this period he saw himself as redefining the scope of pathology in light of the new knowledge to be gained from bacteriology.[26]

Prudden had returned from Europe unconvinced that the Loeffler bacillus was the cause of diphtheria. In *The Story of Bacteria* he wrote that "it is not yet certain whether [diphtheria] is caused by one form of germ, or whether in different outbreaks or in different regions sometimes one and another species induces the disease."[27] In 1889, during an epidemic of laryngeal diphtheria at the Foundling Asylum in New York City, he had an opportunity to conduct his own study on the etiology of diphtheria. He opened his report on this work by noting, as many previous studies had, the difficulties diphtheria posed:

> The thoughtful student of diphtheria is early impressed, alike in his studies by the bedside, at the dissecting table, and in the laboratory, of the indefinite conception which so widely prevails as to the nature and limitations of the disease. The difficulties with which this lack of precision invests the subject are especially felt when he is trying to get light upon its etiology.[28]

In spite of Loeffler's work, Prudden felt his skepticism was justified. He strongly believed that a bacterium was the cause of diphtheria and that knowledge of that bacterium was precisely what was needed to make the "exact distinctions between it and other diseases." Yet he was unconvinced that Loeffler's bacillus had passed the test. He reviewed Loeffler's experiments and concluded that while much had been revealed by this work, "which may be regarded in many respects as models for future workers," Loeffler's conclusions left him unsatisfied. Prudden concluded:

> At present, then, it would seem that actual existing knowledge justifies us only in the conjecture that diphtheria is caused by some form or forms of bacteria. What the species is, or whether there are more than one, we do not know. We are confronted to-day with the same series of questions about which discussion has centered for many years.[29]

Prudden did not set out to offer a complete solution to the problem of the etiology of diphtheria, which he regarded as "one of the most complex and difficult, as it is one of the most important questions in experimental pathology."[30] His goal was to present a series of observations on a number of cases of diphtheria occurring in New York City and to determine the nature of the microorganisms found near the pseudomembrane or in the internal organs with enough frequency to determine whether or not they bore a causal relationship to the disease. In twenty-four cases of clinically diagnosed diphtheria he found, upon autopsy, only the presence of streptococcus bacilli and no Loeffler bacilli.[31] Prudden's results generated some controversy among German reviewers of his work. They read his results as an argument against the etiological importance of Loeffler's bacillus, a contention that Prudden viewed as not justified by his evidence.

The dispute, if it can be called that, between German researchers and Prudden indicates the international scope of diphtheria research in this era. It was concern with the reaction of their German colleagues that spurred researchers in the United States who were most in a position to make an impact in this field — Prudden and William Welch. Just weeks before Prudden's second report on his diphtheria studies were published, William Welch and Alexander Abbott published a short paper on the etiology of diphtheria in response to the German comments.[32]

In 1890 Loeffler had published a widely read critical review of current work on diphtheria, discussing the work of Babes, D'Espine, Von Hoffmann, Ortmann, Roux and Yersin, and Escherich and Klein among others.[33] The evidence he presented seemed to support overwhelmingly the Klebs-Loeffler bacillus, as it was now called, as the causal agent in diphtheria. The other significant finding revealed at this time came from the work of Roux and Yersin in 1888 at the Pasteur Institute.[34]

Roux and Yersin confirmed Loeffler's findings of his bacilli in all cases of diphtheria, but more importantly, they had also demonstrated that the bacilli manufactured a poison. They found this poison in the filtrate of cultures of diphtheria bacilli and demonstrated that when it was injected into guinea pigs, they suffered paralysis and finally death. Their work opened up an entirely new area of research into the nature of this poison. The one discordant note cited by Loeffler in the chorus of affirmations of his findings was Prudden's 1889 paper. Loeffler did not cast aspersions on Prudden's competence as a researcher, though he did note how odd it was that the bacilli that were found regularly by German researchers were constantly missed in the United States. He was confident that further research would clear up the problem: "A simple 'Liquet' can not be affirmed until investigations inconsistent with this, such as those of Prudden in North America, have been shown to be erroneous."[35] To some extent, Loeffler questioned whether what was called diphtheria was the same disease in both countries. This comment could have been a reflection of the problems associated with identifying diphtheria or the distinctiveness of diseases in the United States. This notion had long been a source of debate among physicians here and abroad, and it resonated with older notions about the role of geography in the origins of disease in the prebacteriological era.[36]

Welch, in his growing role as a leader in bacteriological and pathological research, felt that Loeffler's point about the character of diphtheria in the United States deserved a response. Conducting a bacteriological investigation of cases of diphtheria in Baltimore, he and A. C. Abbott quickly identified the source of the problem with Prudden's work. Prudden had taken cultures from children who died in a foundling asylum in which measles and scarlet fever were prevalent. "It seems to us, therefore, clear," they reported, "that the cases examined by Dr. Prudden, taken as a series, should not be regarded as cases of primary diphtheria, but rather diphtheria secondary to scarlatina, measles, erysipelas, or as developing in a situation where these diseases prevailed."[37] They chose for their investigation cases from private practice in which the clinical diagnosis of diphtheria was certain. Examinations were made from bits of membrane taken during life rather than from autopsy. In each case they found the Loeffler bacillus.

They affirmed their belief in Prudden's competence: "We consider that the excellence of his methods, his well known thoroughness and accuracy and his well deserved reputation as a bacteriologist, give to the results published by Prudden the greatest weight." Nor were they interested in arguing about errors in Prudden's conclusions. They recognized that the problem of identifying bacteria in cases of mixed infection of diphtheria and scarlet fever had only recently been noted and that the complete isolation that could have prevented such infections, especially in institutions like asylums, was practically nonexistent.[38]

There are some other results in their study worthy of comment. Welch and Abbott not only examined eight confirmed cases of diphtheria, but they also made bacteriological examinations of cases of suspicious sore throat. These cases consisted of children who had little or no signs of throat infection, along with some suspicious cases of tonsillitis. The bacteriological examination of the cases of tonsillitis showed no Loeffler bacilli. This demonstrated to them the value of bacteriological examination in identifying cases that a physician would have been inclined to view as the onset of diphtheria. Welch and Abbott concluded that their results indeed had confirmed that the Klebs-Loeffler bacillus was present in every case of primary diphtheria they examined. Yet again this report ends with a note of skepticism:

> The only points which can be brought forward as tending to cast any doubt upon the recognition of this bacillus, as the specific cause of diphtheria, are the very exceptional observation of an apparently identical bacillus in the throat of healthy children and the more frequent observation of a bacillus, usually called the pseudodiphtheritic bacillus, which differs from the true bacillus diphtheriae chiefly or only by the absence of pathogenic properties when inoculated in animals.[39]

To Welch and Abbott the presence of the diphtheria bacilli in the throats of healthy persons was an exceptional and puzzling occurrence. They posited that the small numbers of bacilli present in healthy throats indicated that the bacilli did "not find suitable conditions for multiplication and injurious action upon tissues." Therefore some predisposing factor or other morbid condition was needed in addition to the presence of the bacilli to cause diphtheria infection in human beings—for example, the presence of an epidemic, unsanitary conditions, or other diseases. Loeffler, however, argued by this time that the presence of diphtheria in healthy throats was due to the endemic nature of the disease.[40]

In summary, the conclusions of Welch and Abbott pointed to the important questions that all researchers studying diphtheria at this point faced: How was the presence of the diphtheria bacilli in healthy throats to be accounted for? What was the nature of the pseudodiphtheria bacillus? Was it a nonvirulent form of the bacillus, and if so, how and under what conditions did it become virulent? But more interestingly, Prudden, Welch, and Abbott had demonstrated that bacteriological techniques could be used to identify diphtheria in cases that were often clinically defined as laryngitis or a simple sore throat. The "diphtheria" identified by this method, however, thus embraced these other disease entities and was not the same disease defined by clinical observation. The implications of this point would not be revealed until the use of bacteriological diagnosis spread beyond the confines of elite physicians like Welch, Abbott, and Prudden. Their response to Loeffler's challenge served as their entry into the international research on diphtheria. Although none of their work had broken new ground conceptually or technically, they had established themselves in the new field of bacteriology.[41]

William H. Park

William Hallock Park (1863–1939) was one of the first students to take Prudden's two-week course in bacteriology at the old College of Physicians and Surgeons.[42] After completing his medical studies and postgraduate study in Vienna, Park set up a private practice and began a nose and throat practice in clinics at the Bellevue, Manhattan Eye and Ear, Roosevelt, and Vanderbilt Hospitals. In 1890 Prudden offered him a scholarship to study the bacteriology of diphtheria in his laboratory during his spare time.[43]

For two years Park studied diphtheria in Prudden's laboratory, and in July 1892 he published his first paper on his research, "Diphtheria and Allied Pseudo-membranous Inflammations: A Clinical and Bacteriological Study."[44] Park's title is an interesting one. His paper brought both the bacteriological and clinical issues to the fore. It was the first to acknowledge that bacteriology could be integrated with the clinical study of diphtheria. Park examined the questions that all students of diphtheria debated: "Are all cases of diffuse pseudo-membranous inflammation of the upper air-passages, and all cases of membranous laryngitis, rhinitis, and tonsillitis the local manifestation of one disease—diphtheria? Are all cases equally contagious and equally dangerous?" (p. 115). Park was interested in how bacteriology could settle the question of what diphtheria was and what it was not. He was not interested in the debate over whether or not the Klebs-Loeffler bacillus was the specific cause of diphtheria. He believed that previous investigations had settled that question.[45]

Park summarized the current research on diphtheria, citing both Loeffler's 1890 review and the work of Prudden, Welch, and Abbott. He concluded that they "and many others, have established that in all cases of typical infectious diphtheria the Klebs-Loeffler bacilli are present in large numbers in the pseudo-membranes, either alone or associated with other bacteria, and that the Klebs-Loeffler bacilli in all other inflammations of the throat *and in healthy throats are very rarely found*" (p. 115, emphasis mine).

As discussed earlier, the presence of diphtheria bacilli in healthy throats, even in the small numbers of cases, left a cloud of doubt over Loeffler's identification of the bacillus for many years. Park is later credited by his biographer and by historians for the discovery of diphtheria "carriers"—healthy persons who harbored the diphtheria bacillus in their throats. Yet, as this summary indicates, Park was well aware that the presence of diphtheria bacilli in healthy throats had already been observed. His later research would emphasize that this was not a rare occurrence, and his major contribution would be to demonstrate the epidemiological and clinical significance of these carriers.

In his first paper, Park explained the two methods that were generally used to obtain from patients with diphtheria material that could later be examined in a laboratory. If possible, forceps were used to remove a piece of the pseudomembrane without causing damage to the patient's throat. Park's practice differed from this. Instead, he carried with him two glass tubes, one filled one-quarter full with solidified blood serum, and the second containing a sterilized stick about an inch long with a cotton plug wrapped around its end. Using forceps, he rubbed one end of the stick gently against any visible pseudomembrane or against the tonsils or pharynx if no membrane was present. The cotton swab was then lightly rubbed against the surface of the solidified blood serum, then placed back into its original tube, and plugged with cotton.[46] Later he examined the cotton swab in the laboratory. Park carried a number of these tubes around with him as he examined cases of diphtheria, since he found by experiment that he could make reliable cultures from the material collected on the swabs.

Park's kit, known as a "culture tube," provided a simple and effective method for physicians to obtain a specimen from a throat for bacteriological diagnosis. This was not a trivial invention. Welch and Abbott had noted that though the method of bacteriological diagnosis was simple, "it may be questioned whether practitioners of medicine in general are likely to be in a position to make use of this means" because most practitioners had neither easy access to a laboratory nor, in most cases, to the training to take advantage of laboratory findings.[47] A bacteriological diagnosis was an abstract concept for most of them. Park's innovation had the potential to change this. After his kit was made widely available, it would make bacteriological diagnosis visible and viable for the average physician.

Park studied every aspect of diphtheria in the wards of the Willard Parker Hospital. He charted typical symptoms from mild to severe. He examined and characterized the bacilli he found in the throats, noses, and even on the soiled bedclothes of the patients. He found bacilli on soiled linens weeks old. He then focused on how the disease was transmitted from person to person. He related in detail several cases where direct contact involved either another case of diphtheria or infected clothing and bedding. Finally, he noted that those who had extensive pseudomembranes were very sick and that most of these cases died. Those with croupus rhinitis or with the disease confined chiefly to the tonsils and uvula, both children and adults, hardly seemed ill at all.

Park concluded that there were two classes of pseudomembranous inflammations: one caused by the Klebs-Loeffler bacillus and the other by some form of streptococci. The first, "true diphtheria," he described as "a local process, and its lesions are due to the effects of the poison produced by the bacilli in the pseudo-membrane. It is dangerous in all periods of life." The second, "pseudo-diphtheria," he wrote, "is at first a local lesion but may at any time become a general infection. It is peculiarly liable to cause broncho-pneumonia in children." He argued that the name "diphtheria" should be reserved for those cases in which the Klebs-Loeffler bacilli were present "whether alone or associated with other bacteria." He further suggested that cases where streptococcus were more prevalent in the throat should be called "streptococcus diphtheria," but "pseudo-diphtheria" would be preferable since he also included other cocci along with streptococci in his definition.[48]

Park closed with an argument about why the bacteriological examination was so important: it was the only means to establish a correct diagnosis. Without a correct diagnosis, all attempts to learn from statistics the worth of special treatments was useless because of the frequency of incorrect diagnoses. Bacteriological examinations were of great help to prognosis and rational treatment in severe cases and "enable(s) us to take measures more effectually to prevent the spread of the contagion." And finally, bacteriological diagnosis "is certain, can frequently be made immediately and always in twenty-four hours." Park also noted that bacteriological diagnosis would be extremely useful for boards of health:

> As the early detection of diphtheria is important for the general health, and as this disease occurs most frequently among the crowded poor, who are unable to pay for special examination, it would seem peculiarly the business of Boards of Health to undertake it. In small cities some central places could be selected where the necessary appliances could be kept, in large cities several would be necessary. From these laboratories a properly equipped man could be called to make a diagnosis. Children's hospitals and those for infectious diseases should certainly give their pathologist the means to do this.[49]

In his first paper the twenty-nine-year-old Park had made the most comprehensive study of diphtheria in the United States up to that time. He had devised the technical apparatus to easily carry out a bacteriological diagnosis, demonstrated its clinical value, suggested a means to make this knowledge available to a large community of physicians, and shown its potential value for the public health.

Hermann Biggs and the Birth of the Bacteriological Laboratory

Hermann Michael Biggs (1855–1923) was born in Trumansberg, New York. He entered Cornell University in 1879. Like many students interested in medicine at the time, he interrupted his studies for two years to attend medical school at the Bellevue Hospital Medical School in New York City. There he studied with William Welch and Austin Flint Sr. In his last year of medical school, he took a private course in pathological histology with Welch. Upon learning of Koch's discovery of the tuberculosis bacillus, he became interested in bacteriology.[50] He received his bachelor's degree from Cornell in 1882, submitting a senior thesis entitled "Sanitary Regulations and the Duty of the State in Regard to Public Hygiene." In this thesis, which is often cited by scholars as an example of his early views on public health, Biggs asserted that the physical condition of a people was dependent on their sanitary surroundings, which in turn determined their moral and intellectual condition.[51] His early interest in public health placed him within the tradition of middle-class reformers who became more active in late-nineteenth-century New York City and elsewhere. These reformers wanted both to ameliorate the conditions of the poor and the growing ethnic populations and to socialize them to an older Yankee Protestant value system.[52] They saw in public health a vehicle for their purposes. Biggs and other reformers claimed a kind of expertise that they firmly believed could solve the increasingly difficult social problems in the city. By linking sanitary conditions to moral and intellectual conditions, Biggs asserted that public health was critical to social reform and ordered social change. Armed with this vision, Biggs moved quickly to put his ideas into practice with little acknowledgment of the potential conflicts that could arise between medicine and the state.

After graduating first in his class at Bellevue and completing postgraduate work in Berlin, Biggs set up practice in New York City. He held a number of consulting positions at hospitals in the city and also became a member of the New York Academy of Medicine and the New York Pathological Society.

Welch's departure for Baltimore left the new Carnegie Laboratory, which had been created largely for him, without a director. In 1885 the prominent physicians Dr. E. G. Janeway and Dr. Austin Flint were named as directors and Biggs as laboratory instructor. In fact, Biggs took charge of the laboratory, although according to his biographer he had no formal training in bacteriology. In order to prepare for the courses he was to teach at the Carnegie Laboratory, Biggs translated a book on bacteriological methods by the German physician and docent in hygiene and bacteriology at Wiesbaden, Dr. Ferdinand Hueppe.

Dr. Joseph D. Bryant, Commissioner of Health for the City of New York and professor of clinical surgery at Bellevue, enlisted Biggs's expertise in bacteriology in 1887.[53] He asked him to determine the risk posed to the city by large quantities of rags that were being imported from Egypt, where cholera was rampant. He also wanted Biggs to investigate several ships filled with Italian immigrants being held in quarantine in New York Harbor because of suspected cases of cholera. As a result of his involvement in these projects, Biggs approached Bryant with the idea of "establishing a division of bacteriology and disinfection in the New York City Health Department."[54] According to Biggs's biographer, Bryant realized the value of such a department after Biggs and Prudden used bacteriological analysis to establish the presence of cholera on one of the Italian ships in the harbor.

Biggs was to claim later that "Asiatic cholera was then only excluded from

New York City by reason of bacteriological examinations."[55] His assessment was slightly exaggerated. The bacteriological tests only confirmed that the deaths were due to cholera—they did not prevent cholera. Biggs implied that the imposition of quarantine was dependent on the diagnosis of cholera. The diagnosis only helped sustain the quarantine but was not required for its imposition.

In 1888, in the wake of the cholera threat, Bryant, under the direction of Mayor Hewitt, formed a semi-official unsalaried board of consulting pathologists from members of the New York Academy of Medicine's Committee of Conference, which had been established to advise the city during the cholera outbreak. The members included Prudden, Biggs, and Dr. H. P. Loomis. The initial focus of the board was to study the prevention of tuberculosis. This board, led by Bryant, repeatedly but unsuccessfully tried to obtain appropriations from the New York City Board of Health to establish a bacteriological laboratory.[56]

This request reflected the views of Biggs and Prudden in particular, namely that bacteriology was essential for the control of disease, as Koch's work and their own on cholera had shown. Since the control of disease fell under the auspices of boards of health, they felt that municipal and state health departments should establish bacteriological laboratories. Biggs was influenced, as noted before, by his early views on sanitary science and the work of German bacteriologists like Hueppe. Hueppe argued in the book Biggs had translated that as a science, bacteriology could only be put on a firm basis to generate lasting results within special bacteriological institutes that were united with hygienic institutes.[57] He believed that bacteriology "as an object of study . . . would be more practical if combined with hygiene," especially since in his view every practicing physician and medical officer needed to be kept intelligently informed of new investigations in bacteriology.[58]

Prudden had written in some detail about the need for laboratories that would be supported by state or local government boards of health. In 1885 he elaborated his views in his report to the Board of Health of the State of Connecticut on Koch's methods of identifying cholera bacteria.[59] In the conclusion to his report, which was largely a discussion of the specific details of the bacteriological tests used to detect cholera, he discussed how such methods should be applied in the United States. He argued that only the state had the appropriate authority to run such a laboratory, because of its need to contain epidemics. In order to do this most effectively, they needed to have trained bacteriologists. These bacteriologists would be needed to perform a number of tasks: to discover the sources of contagion in disease outbreaks and enforce appropriate measures to destroy them; to perform bacteriological analysis of drinking water and foods; and to conduct original research on the relation of bacteria to disease.[60] Prudden shared with Biggs a sense of urgency that state and local government-supported laboratories — "centres for research and control" — were needed to best utilize the new bacteriological techniques for the prevention of disease.

Biggs's and Prudden's interest in government-funded bacteriological laboratories suggested that they saw a need for an institution that would train and employ people with the expertise they themselves were developing independently. Each served as head of a laboratory and saw the need for the state or some municipal body to address disease prevention systematically in the same way that they were called upon to do on an ad hoc basis from their current positions. Prudden, who did not like to deal with politicians, chose to work behind the scenes; whereas Hermann Biggs moved to make the significance of such an institution and its work known in every public arena.

Although Biggs and Prudden advocated for state and municipal support of bacteriological laboratories, they also knew that there were problems with government sponsorship. Specifically, state and city governments were controlled by politicians, not by physicians or sanitarians. The New York City medical community had long decried the fact that the president of the Board of Health was not a physician. Physicians had long felt that a lay president reflected poorly on the medical profession, which was working steadily to improve its image and increase its authority over health matters in the city. Machine politics, specifically the machinations of Tammany Hall, dominated political life in New York City in the 1890s. Reformers regularly complained that too many positions in city government were dispensed as patronage to Tammany supporters rather than filled on the basis of merit. The bacteriological laboratory in the New York City Health Department was created in the fall of 1892 after a summer when relations between physicians and the department reached a particularly low point. The accusation that recent personnel changes in the health department were politically motivated drew a great deal of press.

On June 25 several of the members of the Medical Consulting Board to the Board of Health resigned. In a fairly unusual move, two members of the board, Abraham Jacobi and T. Mitchell Prudden, publicly stated that they were resigning because the Board of Health had fallen into the hands of

Tammany Hall.[61] The president of the Board of Health, Charles Wilson, who was not a physician, denied that any outside pressure had been placed on him and claimed that he had replaced Ewing as sanitary superintendent and demoted the registrar of records solely for the good of the department.[62] Physicians in the city were unconvinced.

On June 29 prominent physician and sanitarian Dr. Stephen Smith also resigned his consulting position to the department, noting that the department had violated its agreement with the advisory board:

> Politics was to be absolutely avoided in order that the best professional service could be fearlessly rendered to the city. . . . Eventually a change was made by which a layman became eligible to the presidency of the Board. This naturally led to a politician being placed in that office and to a department being turned into a part of the patronage dispensing machine apparently inseparable from political work.[63]

By July 6, 1892, the remaining members of the Medical Consulting Board had resigned. The necessity of this act confirmed for many of the city's elite physicians that medical interests and political interests in the city could not be resolved and that public health work would continue to be tainted by politics.

The membership of the two leading organizations of physicians in the city, the County Medical Society and the New York Academy of Medicine, had long been concerned that if they had close relations with the health department, the profession would be tainted by machine politics. Health Commissioner Bryant had worked hard during his tenure to improve relations between physicians and the health department and had initiated the Medical Consulting Board in order to gain more physician support for public health efforts.[64] Yet in the wake of the resignations, Bryant's hard work seemed for naught. Two of the leading newspapers published scathing editorials. The New York Tribune asserted, "Tammany's relentless pursuit of spoils has seldom been more strikingly and repulsively illustrated than in the recent manipulation of the Health Board."[65] The New York Times emphasized the potentially long-term effects of the scandal: "The Board of Health cannot be used as a political machine in the service of Tammany Hall, or managed under the sway of dictation of politicians, without driving from its assistance all physicians of high standing in the community."[66]

Meanwhile, the leaders of New York's reform political groups, who were interested in overthrowing the power that Tammany held over the municipal government, used the resignations of the Medical Advisory Board in their public campaign. On August 13, 1892, the City Reform Club announced that it was sending out 25,000 to 30,000 copies of a leaflet entitled "Politics in the Health Department." The text of the leaflet cited the recent firings and the resignations of the consulting board. It also contained a reprint from an editorial from the New York Medical Journal attacking President Wilson. The editors wrote, "Any Board of Health in this city or any other that has for its head a mere politician is like a pyramid built upon its apex instead of its base."[67]

The timing of the resignations and the Reform Club's use of it may have had to do with the broader political battles in the city. Historian Gordon Atkins has suggested that since the story of the political attack on the Board of Health did not appear until the middle and end of June and the first week of July, when the Democratic National Convention was in session, the resignations may have been timed to take advantage of this. The editors of the New York Medical Journal subsequently printed a retraction of its editorial attacking Wilson and acknowledging mistakes in their reporting of Board of Health statistics. But they held to one point: "That the president of the Board of Health is a politician is perhaps unavoidable. It is nevertheless to be regretted." While the Reform Club's efforts drew attention to politics in the health department, they did not result in overturning the personnel changes.

Though party politics may have contributed to the timing of the resignations, there was frustration among physicians nonetheless. By the middle of August 1892, all the members of the Medical Consulting Board except for Dr. George Shrady had resigned their posts, and Prudden had resigned from his position as consulting pathologist as well. Responding privately to a request from the City Reform Club about his reasons for resigning, he wrote: "A general and growing distrust of the methods of the Health Board under its present head and his wardens in those departments which I had occasion to know most about . . . [has] led me for some time to wish to free myself from all further association with it."[68] Another physician, in an anonymous letter to the editor of the New York Medical Journal also acknowledged the politics in the health department: "It has been an open secret for years that appointment as surgeons to the police force (notwithstanding the examination), to the various medical positions in the board of health and to the nonremunerative positions as members of the visiting staff of the various municipal hospitals were matters of political influence."[69]

That Health Commissioner Bryant and Hermann Biggs did not resign their posts during the summer of 1892 seems unusual given their close ties to the men who had stepped down. Bryant perhaps remained on the board due to his secure position in the health department and the fact that he had survived other attacks from Tammany.[70] This episode is an example of Biggs's ability to maintain his base in the health department despite the machinations of Tammany Hall and also illuminates the troubled relationship between physicians and politicians in the city.

Establishment of the Division of Pathology, Bacteriology and Disinfection

As a cholera epidemic spread around the world in 1892, Biggs, in his role as consulting pathologist and bacteriologist to the New York Quarantine Station, went to Europe to investigate quarantine procedures and the status of medical inspection of immigrants at ports where cholera was likely to strike.[71] In late summer 1892, cholera broke out in Hamburg, Germany's big port city. New Yorkers, living in the nation's busiest port city, were justifiably alarmed.

As Charles Rosenberg has written, cholera was the classic epidemic disease of the nineteenth century as plague had been for the fourteenth century. Its defeat was a reflection not only of progress in medical knowledge and philosophy but of enduring changes in American social thought.[72] Although cholera had not appeared in the United States since 1873, its presence in other parts of the world, the havoc it provoked, and the erosion of the protection offered by physical distances in the wake of modern methods of transportation created a sense of hypervigilance on the part of sanitarians. Furthermore, as a result of Koch's discovery of the bacillus that caused it, cholera could only take hold in those places where officials were unwilling or incapable of using the methods of modern science to contain it. Responding to the cholera scare in the late summer and fall of 1892, the New York City Board of Health established a bacteriological laboratory within the new Division of Pathology, Bacteriology and Disinfection.

Many writers have made the point that New York City's bacteriological laboratory was created in the fall of 1892 because of the threat of cholera to the city and because it was widely believed that bacteriology protected the city from an epidemic.[73] However, this view fails to explain why the Board of

Health did not respond as it had in the past with an ad hoc group of advisors and physicians who would address the immediate threat of cholera and then disband once the threat had passed. In other words, the question remains why a permanent institution was established. It is plausible that the Board of Health was responding not just to the specific threat of cholera that occurred in 1892 but, more important, to the *continuing* threat of cholera. No one believed at the end of the scare that cholera would not appear in New York Harbor at some point in the future. In addition, Biggs moved quickly to show the Board of Health that a permanent bacteriological laboratory would be necessary for the control of other diseases as well.

Cholera appeared in New York just as the health department was under severe attack from physicians and political groups. Its credibility as a department that could effectively protect the city's health in a crisis was questionable. As reports of cholera's movement across Central Asia into Europe increased over the summer, the New York City Board of Health began preparations to act should an infected vessel arrive in the city's harbor. There appeared to be little debate that a rigid system of quarantine and inspection of arriving ships was the best way to protect the city. New Yorkers were assured that such a quarantine would protect them. After the first reports came in from Hamburg in August, George Shrady wrote to the city's physicians, as he had throughout the summer, "There is, as we have said before, no ground for alarm, even should a few cases reach quarantine."[74]

When the steamship *Moravia* arrived on August 30, 1892, from Hamburg with news that twenty-two passengers had been buried at sea after succumbing to severe intestinal attacks, city leaders immediately announced to the public that everything was being done to protect the city. On August 31 Biggs was asked to conduct bacteriological examinations to determine the cause of the deaths on the *Moravia*. Meanwhile, the Chamber of Commerce appointed a committee to evaluate the quarantine situation in the harbor. The committee consisted of members of the Committee of Conference of the New York Academy of Medicine. The members of this committee were the very physicians, with two exceptions, who had severed their connection with the health department in June. It may be that those who had resigned were nonetheless willing to work with the city's politicians in a time of crisis though not on an ongoing basis.

The management of the quarantine was garnering a great deal of public attention and dismay. News reports noted that cabin-class passengers were being forced to stay on board ships where cholera cases were still occurring.[75]

President Wilson daily assured the city that the health officer of the port, William Jenkins, had everything under control. To defend the department from any criticisms, Wilson made sure that other prominent physicians and sanitarians reviewed the work on the islands in the harbor. Yet the quarantine situation worsened. Facilities on Hoffman Island, where steerage passengers were held and their belongings disinfected, had rapidly become overcrowded. Furthermore, there was no place to put the growing number of cabin passengers. By mid-September, Welch received an urgent telegram from Health Officer Jenkins requesting his help. Welch made a hasty visit to New York but had little to offer except to encourage Jenkins to seek Prudden's advice.[76]

The arrival of the *Moravia* was followed on September 3 by that of two other vessels from Hamburg, the *Normannia* and the *Rugia*. Each reported deaths at sea from "cholerine," a catchall term used to describe intestinal illness resembling cholera. Often cases of cholerine turned out to be true Asiatic cholera. New Yorkers began to realize that the quarantine might not hold and that cholera might then enter the city. President Wilson issued the first of his daily cholera bulletins on September 6. He told New Yorkers that "should the germs find their way into the city through another channel, then we are ready for them." He then asked Dr. Cyrus Edson, the sanitary superintendent, to make sure that inspectors of food were especially vigilant. Wilson claimed that up to September, as much as "687,848 pounds of meat, fruit, vegetables and milk [determined to be] dangerous to health had been destroyed." The Board of Estimate, which controlled the city's finances, granted additional funds to extend the work of the summer corps of physicians who served as inspectors of contagious diseases in the city. Edson had been ordered to divide the tenement districts of the city into "cholera districts" with inspectors, nurses, and disinfectors assigned to track down and report every case of persons suffering from intestinal diseases in these areas. Wells, sewers, water closets, and defective hydrants were overhauled throughout the city as well.[77]

On September 9 Bryant went before the Board of Estimate to request funding for the establishment of the Division of Pathology, Bacteriology and Disinfection with Biggs as its head. The request was granted. Six days later, President Wilson invited the New York Academy of Medicine's Committee of Conference to evaluate the preparations within the city for the treatment of cholera patients and to suggest measures to prevent the spread of the disease within the city. On this same day, Prudden was also reappointed to a consulting position in the health department. This time he was to consult specifically to the new Division of Pathology, Bacteriology and Disinfection.[78]

By the end of September eleven cases, including six deaths from Asiatic cholera, had been reported in the city. Biggs and Dr. Edward K. Dunham had been responsible for making the bacteriological examinations of these cases, which they did at the Carnegie Laboratory since, despite Wilson's assertions, the new city laboratory was not yet equipped to carry out the tests. All of the cases were examined after death. Before Biggs announced the cause of these deaths, Prudden was called in to examine the cultures. Samples of the cultures were also sent to Dr. Harold C. Ernst at Harvard University Medical School and to Dr. Petri of the Imperial Board of Health in Berlin. All confirmed the diagnosis of cholera.[79]

No one knew how the cholera had gotten into the city. Biggs, along with one of the inspectors, searched for the source of infection for each case. They could not reach a conclusion; the cases were scattered across the city, and there seemed to be no connection between them. Biggs could only surmise that many mild cases of cholera in the city had not been recognized.

By the end of September no other cases were reported, and it was evident that an epidemic had been prevented. Wilson claimed that the city's health department had done everything that could be done to protect the city during the current scare. In addition, he asserted that should the cholera appear at some other time, "all that science can do, has been done in the way of preparation for the work should the pest come; all that science can suggest to lessen the evil effects of the pest, should it break out, is either finished or now in the course of completion."[80] He viewed the new division of bacteriology and disinfection as a necessary and permanent safeguard against cholera.

Many reports confirmed that New Yorkers did not want to be "like Hamburg," that is, failing to employ all modern, scientific means to prevent cholera. Like others, Wilson was dismayed by the reports from Europe throughout the summer that "cholerine" was present in these cities, but not true Asiatic cholera. The New York health authorities widely regarded this as a ruse and an attempt on the part of these cities to avoid the rigid quarantine that a report of cholera would bring. The health authorities of Hamburg were seen to be the most negligent in this regard. As one writer noted, "They have displayed a degree of criminal inefficiency, negligence, and deceit in dealing with the cholera, that is almost incredible. For two weeks the disease was in the city, the number of victims constantly increasing, before any warning was given, and even then it was said to be only cholerine, and of very little import."[81] Wilson and other New York City leaders had been made aware by the example of Hamburg and the entreaties of Bryant, Biggs, Prudden, and others that bacteriological examinations provided the critical information needed to identify suspicious cases of intestinal disease when the threat of cholera was imminent and the possibilities of the failure of quarantine so vast.

The threat of cholera to New York did not end with the scare in 1892. During the following summer in 1893, reports of cholera outbreaks appeared in the press. And in the view of some, as long as immigrants came in to New York there would be cholera. The editor of the *Medical Record*, George Shrady, wrote in January 1893, "Immigration is accountable for the introduction of cholera into this country."[82] In the fall of 1893 he added, "Considering the continued prevalence of the disease in nearly every country of Europe, its ravages in Russia, Egypt, and Arabia, there is every reason for the exercise of vigilance on the part of the authorities."[83] The continuing threat of cholera, coupled with the desire of the health department leaders to be viewed as responsible managers of the city's health, thus provided support for the establishment of a division for bacteriology and disinfection.

The health department received favorable reviews for its handling of the 1892 cholera scare. The creation of the Division of Pathology, Bacteriology and Disinfection was applauded, along with the choice of Biggs to head it. In an editorial in the *Medical Record* in January 1893, George Shrady claimed that the bacteriological work of Biggs and Prudden, among others, had saved the city from cholera in 1887 and 1892.[84] One physician offered a more vivid assessment: "The days were lurid; quarantine was not a science; cholera was down the bay; Biggs and Prudden stood at the gate."[85] In subsequent reports of the cholera outbreak, there is some question as to how much credit should be given to bacteriological analysis in preventing cholera from entering the city.

Biggs, as might be expected, gave great credit to bacteriological analysis in his report on the outbreak written a year later. He noted that some (unnamed) persons questioned the importance placed on bacteriological examinations in protecting the city:

It would seem as if an argument to establish the necessity of such examinations in the beginning of an outbreak of Asiatic cholera was on a par with an argument to show the importance of the examination of the sputum in assisting in the diagnosis of doubtful cases of tuberculosis. So far as I am aware, in the absence of an epidemic or proof of infection, all physicians of experience are agreed that it is impossible to make, with certainty, a clinical diagnosis of Asiatic cholera. . . . I greatly question whether credence would or should be given to the statement of any physician or any number of physicians in asserting, upon the clinical history alone, that any given isolated case, occurring in the absence of any proof of infection of epidemic cholera, was a case of that disease.[86]

Though Biggs's point displays his passion and commitment to bacteriology, not everyone believed that bacteriology had saved the day in this particular case. As discussions evaluating the handling of the cholera outbreak confirm, most observers believed that the quarantine, the vigilance of the health inspectors, and other measures taken to make sure the city was clean had the greatest impact in preventing an epidemic.[87] Given the fear that cholera aroused, physicians argued that the presence of cholera could be determined by the clinical history and postmortem appearances. "To avoid a senseless panic," Dr. Alfred Carroll told the members of the New York Academy of Medicine, "it should be known to a certainty whether one had to deal with Asiatic cholera, and this could now, or in the past, be determined by the clinical history and postmortem appearances without waiting several days for a bacteriological examination."[88] Bacteriology was not dismissed by physicians and sanitarians in the city. Indeed the early diagnosis of the disease by bacteriological tests was viewed as a necessary part of the scientific management of cholera. Still it was not seen as more important than disinfection, isolation, or quarantine.

It appears that the efforts of the health department to enlist physicians in protecting the city against cholera had helped to heal the breach of August. Although the city actively sought help from the medical community, some physicians remained skeptical that the department could ever completely remain above politics. The Medical Advisory Board was not reestablished by the end of 1892.

Prudden emerged from the summer scandal and the cholera scare with a renewed sense that a national bureau of health was essential. He insisted that such a bureau had to be staffed with trained, experienced men chosen for their special knowledge, not from "the flotsam and jetsam of the political ocean, from which too often strange, uncouth things are stranded in offices where malfeasance may mean death to some, disease to many."[89] Prudden remained committed to the use of bacteriology in the service of public health, yet he was now convinced that only the national government could promote public health interests and ensure that local politics could not interfere.

Prudden's and Biggs's dream of a bacteriological laboratory for the city was realized in the fall of 1892. While they anticipated some of the problems they would face as medical men seeking to apply scientific knowledge to issues of public health in an institution not totally under their control, there is little evidence that they foresaw how complex the relationships between the city and physicians would continue to be in the long term.

Bacteriological Diagnosis of Diphtheria

The new Division of Pathology, Bacteriology and Disinfection of the New York City Health Department, established in the fall of 1892, consisted of nine assistant medical inspectors and sixteen foremen and laborers. Two small rooms on Bleecker Street near the headquarters of the health department were set up as a laboratory. In January 1893 an additional twelve medical inspectors and a number of lay disinfectors were added to the staff.[90]

The first program taken up by the new division was the bacteriological diagnosis of diphtheria. The program was first put forth by Biggs in January, when he made two recommendations to the Board of Health: first, that they begin bacteriological examinations for the diagnosis of diphtheria, and second, that William Park be appointed as a special inspector for this work.[91] Biggs recommended Park because he was familiar with his diphtheria investigations.

Biggs's argument for the new program to the Board of Health was a shrewd one. He laid out the benefits of his program in terms that were very specific to the situation in New York City. First, he noted that the differentiation of diphtheria from other inflammations of the throat was of great sanitary importance because of the large number of deaths attributed to these diseases in the city. (In 1891 there had been 4,874 cases and 1,361 deaths from diphtheria in the city—up from 1,267 deaths in 1890.)[92] Second, he emphasized that all clinicians recognized that it was impossible from the clinical history or the anatomical lesions or from both to make an accurate diagnosis of diphtheria. He cited Park's bacteriological examinations of diphtheria cases in the city's own Willard Parker Hospital, which showed that from 30 to 50 percent of the patients did not have true diphtheria. Those cases that were not true diphtheria posed little danger to others; whereas cases of true diphtheria, even those of the mildest type, were known to generate other cases of diphtheria. Most importantly, Park had demonstrated that an accurate diagnosis of diphtheria could be made by bacteriological examination within twelve to twenty-four hours. Third, he pointed out that the economic savings promised by the use of bacteriological diagnosis were potentially quite significant since diphtheria appeared to be increasing. Under his plan no unnecessary disinfections would be carried out, and only cases of true diphtheria would be monitored and removed to the hospitals. Fourth, Biggs appealed to the city officials' sense of pride. The use of bacteriological diagnosis, he asserted, would be an important advance in public health practices. Biggs even went so far as to claim that New York had made the commitment to rely solely on bacteriological examination for the diagnosis of cholera in 1892. No other state or municipal board had officially adopted the use of bacteriological diagnosis in diphtheria as the city had with cholera. Yet, he asserted, the diagnosis of diphtheria was of greater importance in New York City than cholera because of its greater prevalence and its constant presence. Sanitarians would hail such a program as a sign of the New York Board of Health's determination to employ the most recent advances of scientific medicine. Finally, Biggs promised that this program would also promote investigations into the best methods of diphtheria prevention.[93]

By linking diphtheria to the known and terrifying threat of cholera, Biggs was able to make the more constant, yet less alarming disease more visible to a Board of Health that was most probably unaware of the specifics of its prevalence in the city. Biggs's recommendations also hinted that some great recognition was in the offing for the Board of Health if it took the chance and supported the program he proposed. Bacteriology was the key, he implied, to increasing the status of the health department. His argument was persuasive. After some delay, the Board of Health approved Biggs' request. In May William Park was appointed as inspector and bacteriological diagnostician of diphtheria. Simultaneously, bacteriological examinations for the diagnosis of diphtheria were ordered for all cases admitted to the city hospitals.

Between January, when Biggs first made his recommendations, and May, when Park was hired, physicians discussed the merits of bacteriological diagnosis. An editorial in the *Medical Record* noted that while more and more physicians believed that the Klebs-Loeffler bacillus was the cause of diphtheria, questions remained about its value in distinguishing diphtheria from other throat infections. Bacteriological examinations were potentially useful, the editorial argued, but largely impractical for the average physician: "It is of course, rather unfortunate that the immediate result of these observations cannot give much assistance to the practitioner who is in doubt as to the true nature of suspicious cases, for the physician must act at once, and will be unable to await the result of a bacteriological examination."[94] Other physicians argued that bacteriology had only identified the specific microbe said to cause diphtheria and that this information had little impact on what the physician had to do to effectively treat the disease.

In February 1893 Park published his second paper on diphtheria. In this paper he extended his previous investigations with the intention of simplifying his methods and shortening the time needed to make a bacteriological diagnosis, which he recognized would make it of more practical use to the physician. The paper addressed the kinds of questions that caused concern among physicians. Park restated his categorization of diphtheria and pseudodiphtheria based on the presence of the Klebs-Loeffler bacillus. He confirmed that the bacillus was found in both mild and severe cases. "The absence or slight development of the pseudomembrane, and the few symptoms in some of the cases in which the Loeffler bacilli were found," he wrote, "have caused some to deny they were diphtheria. We have clinical as well as bacteriological evidence that these mild cases are true diphtheria."[95]

Park reported on several cases in detail to illustrate his points, including cases of adults with mild symptoms who had been in contact with children with diphtheria, cases of children in schools where other cases had occurred, and cases displaying a range of symptoms in an epidemic that occurred on board a Danish ship in New York Harbor. He also described typical cases of pseudodiphtheria caused by streptococci and other bacteria. Park demonstrated in this paper the ways in which the use of bacteriological examinations could provide the information physicians needed in evaluating all cases of diphtheria from the difficult-to-diagnose mild cases to the most severe cases. He explained the process of making the bacteriological diagnosis, describing the necessary tools and how to use them properly. What was perhaps most persuasive about his argument was the way he took the reader through each step by putting himself in the physician's place and walking through the process to obtain a diagnosis of diphtheria. By taking this approach he showed skeptical physicians its value and thus made it plain how it was their "duty" to make a diagnosis using cultures.

In June 1893 the Board of Health announced that it would provide bacteriological diagnoses of diphtheria to all physicians in the city. Thirty-four depots were established across the city, usually in drugstores, where physicians could pick up and deposit culture tubes. Health department workers collected the tubes late in the afternoon each day and a report was available to the physician by noon the next day. The physician could call in and get the report immediately; otherwise it was mailed.[96]

The culture outfit consisted of a small wooden box containing a culture tube, swabs, and a blank for recording the name, address, and other information about the patient. The instructions also directed physicians to give a preliminary assessment of the patient's symptoms, the duration of the disease, the location of the membrane, and a clinical diagnosis. A circular accompanying the kit described the difference between true and false diphtheria and, more important, asserted that the difference between the two could only be determined by bacteriological examination.

The circular also emphasized that the health department wanted physicians themselves to take the material for cultures of their patients. The department's medical inspectors would intervene only at the attending physician's request. This point intentionally addressed a growing tension that the bacteriological program had produced. While most physicians initially responded positively to the health department's diphtheria diagnostic program, some questioned the status of their authority as physicians versus that of the health department in making a diagnosis. "It is surely a most radical proposition for the State to step in and offer to become the diagnostician of disease," the *Medical Record* noted. "We shall watch the experiment with great interest."[97] The editors of the *New York Medical Journal* recognized that the health department was taking a radical step, yet they voiced support for the program, given that it left diagnosis in physicians' hands:

> *Heralding the assumption of what appears to be a new function in boards of health—that of giving to the individual physician the aid of resources ordinarily beyond his reach and far more acceptable to him than an inspector's visit to his patient . . . In view of the well-known skill of the board's bacteriologists, it is to be taken for granted that the light swabbing specified in the circular is considered sufficient for*

the purpose; they are not men to take upon themselves the responsibility of declaring a diagnosis—for practically that is what their report amounts to—without all the necessary data attainable.[98]

These editors acknowledged Biggs's and Park's expertise and the potential value of bacteriological examination to physicians. They lauded its use, while reserving comment on whether or not the new role the health department had claimed for itself as diagnostician was a problematic one. In his second paper, Park had tried to make it clear that physicians could take material for the cultures themselves as he had done. The expense associated with the equipment needed to make the kit was not prohibitive. He described how to make the blood serum and the swab, and even indicated where materials could be purchased.[99] He also carefully described the laboratory techniques used to grow the cultures and examine the bacteria, though it is clear he did not expect physicians to carry out this aspect of the work. Before he joined the health department, Park gave no indication that someone other than the physician would or should perform the necessary tasks to obtain the material for a bacteriological diagnosis of diphtheria. He had argued that the health department should take on the function of using bacteriological diagnosis *only* among the poorer classes. The health department had been careful to emphasize in its circular accompanying the culture kit that the department's inspectors would not interfere with cases under a physician's care. Yet it was readily apparent that, in the department's view, bacteriological diagnosis was the only acceptable diagnosis of diphtheria, and physicians would have little room to dispute it.

By November 1893 it was clear that some difficulty had arisen in the application of Biggs's program with respect to the role of clinical diagnosis in diphtheria. In a letter to the editor of the *New York Medical Journal*, Dr. Stuyvesant Morris complained that shortly after he reported a case of diphtheria in a private house, a health department inspector visited the family. The inspector demanded to see the patient in order to obtain a culture, saying to the mother of the child, "If the physician knew his business, he would have sent them a specimen." Dr. Morris was outraged:

I have written twice to the department asking by what right such a demand was made. . . . Has there been any recent enactment which gives them such right? No offer was made to examine the sanitary condition of the premises, or to find out what precautions were being taken for the protection of the other inmates. . . . If there had been any question of diagnosis I would have taken a culture for my own satisfaction. In their zeal to obtain this, it seems to me they overlooked matters more important than the diagnosis.[100]

While this case may have been an example of an overzealous inspector, other physicians complained as well. A homeopathic practitioner in Brooklyn protested against the "unwarranted interference with private practice."[101] The following year an article in the *British Medical Journal* pointed to a potentially more serious problem with the program. It stated that since the bacteriological diagnosis was used so generally in New York City, the clinical diagnosis of diphtheria "is likely to occupy a less important position than it deserves."[102]

These two issues, the usurpation of the role of diagnostician by the health department and the potential for the subordination of clinical diagnosis to bacteriological diagnosis emerged as the most problematic aspects of the diphtheria program. Tensions had long existed in New York City between physicians and the medical inspectors employed by the health department. In the highly competitive New York medical market, paying patients were at a premium.[103] Young physicians found it difficult to establish a private practice that could support them in a middle-class lifestyle. Many of them augmented their private practice by working in part-time positions as inspectors for the health department. The large ethnic poor and working-class populations of the city represented contested territory among physicians. And it was here that the health department exercised its fullest authority. Though many among the population were poor, the boundaries between the poor and working classes were not always clearly discernible. Sometimes health department inspectors imposed their authority on patients who, though living in a poor district, did indeed have a private physician available to them. By mandating the use of bacteriological diagnosis, physicians working under the auspices of the health department could further undermine other physicians' relationships with their patients by suggesting that clinical diagnosis was less reliable.

The authority claimed by the health department with respect to diphtheria rested in part on the assumption that bacteriological diagnosis left no doubt about the distinction between true and false diphtheria. Biggs was very clear about this when, in November 1893, he reported on the diphtheria program at a meeting of the Practitioner's Society of New York.[104] He began by reviewing the reasons why he believed the department should take up the bacteriological diagnosis of diphtheria. He had been impressed by Park's work at the Willard Parker Hospital. Especially significant to him were findings that in the cases where the Loeffler bacillus was not found, the mortality was almost nil; whereas in cases where the bacillus was present, mortality was very high—running from 25 percent to 40 percent and even up to 50 percent. Biggs was asked if the presence of the Loeffler bacillus in a case of sore throat would be sufficient reason for placing the patient in a ward among others with diphtheria. Biggs responded that there could be no doubt of it. He continued:

> The bacteriological examination and the clinical history tally in nearly every case. Also, where the Loeffler bacillus is found there were apt to be other cases; where it is not found the disease was likely to be limited to the single case. In one of the worse sore throats I have ever seen diphtheria was excluded by the absence of the Loeffler bacillus, and the patient got well.[105]

Biggs was not willing to admit that there was any discrepancy between the clinical and bacteriological diagnosis of diphtheria or that the bacteriological diagnosis presented any problem. He was so confident of this position that he was prepared to exercise the full police power of the health department when challenged (and had done so).[106] When asked how the public had responded to the practice of removing diphtheria cases based on a bacteriological diagnosis, he replied that the inspectors had met with little trouble, "but when opposition was offered it usually ceased when one of the sanitary police accompanied the ambulance."[107] Biggs felt justified in this use of force because the bacteriological program had indirectly revealed that diphtheria was very prevalent in the city, and thus, in his view, extreme methods were warranted. Here Biggs made one of his first elaborations of the link between scientific evidence as justification for the fullest expression of state power with respect to public health.

When bacteriological examinations were extended to all cases of suspected diphtheria occurring in tenement houses and boarding houses, the department gained some valuable new information. In a great number of cases, children convalescing from diphtheria still showed the presence of the Loeffler bacillus even after the membrane had disappeared from the throat. In some cases they found the bacillus for up to five weeks. As a result, the health department mandated that all cases of diphtheria should not be discharged from the hospital or allowed to return to school until a second culture showed the absence of Loeffler bacillus.

Second, the bacteriological program provided the department with a more accurate picture of the distribution of diphtheria in the city. By plotting each case on a map of the city, they determined the centers of infection located across the city. For example, in August four cases had occurred in a house on East 73rd Street. Within a month twelve cases of diphtheria had been reported within a radius of 100 feet of this house.[108] This epidemiological evidence supported the department's increasingly aggressive efforts in tenement districts. As Biggs noted, with this evidence the health department was ready to remove even more people to the hospital than they had previously, especially from "localities where apartments were small, the number of children in the house large; where parents are ignorant or negligent and where it was impossible to do anything in the way of isolation. Patients would be removed forcibly if necessary." There is little evidence that physicians objected to Biggs's program to control diphtheria in the tenement districts, nor is it clear that he was asked where the additional patients would be placed, since facilities at the Willard Parker Hospital had not been expanded. In this instance, physicians displayed little interest in the exercise of the health department's police powers when it came to the removal of the poor from their homes.

Biggs's assertion that the bacteriological diagnosis concurred with the clinical diagnosis in every case was challenged by physicians in the city. In a letter to the *Medical Record*, Dr. Brendon raised a point about the accuracy of the test. He had taken cultures from two cases in which there "could be no shadow of a doubt regarding diagnosis." The lab report for one reported no exact bacteriological diagnosis, since the culture was made too late in the disease. The report on the second case claimed that the bacilli present only slightly resembled diphtheria, and a second culture was requested. The child died before the second culture could be made. The doctor asserted that both cases showed the typical local and constitutional symptoms of diphtheria. Troubled by this discrepancy, he posed this question to the New York medical community:

> If we cannot get a correct and definite report in a typical case, how can we rely on the
> report in a doubtful case? I have frequently had reports on cultures from the Health
> Department laboratory where the clinical symptoms did not bear out the bacteriolog-
> ical report but this is the first time it has been clearly demonstrated to me that the
> report is a slim reed to lean on, and for this reason I still further emphasize my objec-

tion to strange doctors taking cultures from every case as a routine procedure. I may
say that in conversation I have heard several medical neighbors make the same com-
plaint of interference and unreliability of report.[109]

As Dr. Brendon's letter indicates, physicians also saw their authority to make
a diagnosis being challenged by the diphtheria program. While they did not
seem to register objections regarding tenement districts, private patients
were another matter. In these cases, both the authority that a doctor gained
from accurately characterizing a patient's disease and other aspects of the
doctor-patient relationship were being disrupted by health department inspectors, those "strange doctors," as Dr. Brendon put it.

Diagnosis is probably the most fundamental task in the practice of medicine. In an era when standards of medical practice and education were minimal and competition for patients fierce among regular and homeopathic
practitioners, the doctor was primarily dependent on the personal relationship he maintained with his patients.[110] The use of results from a bacteriological examination that he did not control put the physician in a tenuous
position with respect to his patients should that report not confirm the doctor's or the patient's own views.

But while Dr. Brendon had raised an important point about the accuracy
of the laboratory report, his argument was undermined because he failed to
understand a crucial aspect of the bacteriological diagnosis of diphtheria:
there was no "typical" case of diphtheria as defined by clinical symptoms. As
Park and Biggs had argued, no certain diagnosis of diphtheria could be made
based on the clinical symptoms alone, no matter how mild or severe. Dr.
Brendon was also confused about other aspects of the culture tests, namely
the need for time and skill.

While Park could claim with confidence that the use of cultures provided
the most reliable information to make a diagnosis of diphtheria, he had been
careful to point out some caveats. First, the culture was most useful when
taken early in the disease. When taken at a later point, "the presence of the
Klebs-Loeffler bacilli in the cultures makes the diagnosis certain, but the absence of the bacilli does not prove that the case had not been diphtheria."[111]
He had found that the Klebs-Loeffler bacilli tended to disappear after the
fifth day; therefore, the amount of assurance that the culture test provided
was time-limited. The form that the health department sent out with the
culture kit did not explicitly note this fact, though the physician was asked

to furnish information about the duration of the disease at the time of the culture.

The second caveat was that a certain amount of skill was required to take a proper culture. Biggs had noted that in the beginning of the program "some physicians familiar with bacteriological work feared it was unwise to trust the inoculation of the culture tubes to physicians unskilled in bacteriological methods."[112] After the program was initiated, Biggs had concluded that physicians could be relied on to take proper cultures. But this point was disputed to some extent by Park and Alfred Beebe in their much more comprehensive report on bacteriological diagnosis, which appeared in the Annual Report for 1893:

> Many physicians, as well as the inspectors, gradually became so skilled in making inoculations that it was possible to rely certainly on the results obtained from the examination of their cultures, while, on the other hand, it was found that caution was necessary in accepting the inoculations of others, and in such cases a second culture was requested.[113]

Park and Beebe had also found that physicians did not always know or remember that cultures taken after irrigating the throat with antiseptic solutions were likely to be unreliable as well. Thus to some extent the reliability of the bacteriological diagnosis was dependent on a number of conditions that only the physician taking the culture could control. When the inoculations were poorly made or faulty in any way, the diagnosis was always suspect. The informational circular dispensed along with the culture kit urged physicians to follow the instructions carefully. But it did not make clear how the lack of care could undermine the diagnosis they received from the health department.

Problems with the interpretation of a bacteriological diagnosis continued even among those engaged in making the examinations. In April 1894, almost a year after the diphtheria program began, Dr. Walter Chappell raised some important questions about the work. Chappell was traveling around the city with Park, making bacteriological examinations and attempting to settle the question as to the contagious or noncontagious nature of tonsillitis. He examined cultures from forty-seven children in the New York Orthopedic Hospital. Numerous Klebs-Loeffler bacilli were found in two cases, and fewer bacilli in two other cases. Since there was no diphtheria in the hospital, Park expressed doubts about the virulent character of the bacilli. After conducting virulence tests on guinea pigs, he found them to be nonvirulent.

Chappell was disturbed by this result. Until this episode, he had understood that finding the Klebs-Loeffler bacilli in a sore throat meant diphtheria, whether or not the clinical symptoms supported the diagnosis. "If we have to rely on inoculation for the final test as to the true character of the bacilli, the simple finding of Klebs-Loeffler would not be available for reliable and rapid diagnosis," he pointed out, "as non-virulent Klebs-Loeffler might be found in a person suffering from follicular tonsillitis." Chappell also saw other troubling implications in this finding. He asked,

> Does the inoculation of an animal with the Klebs-Loeffler bacillus, producing virulent results, necessarily prove that the bacillus is the cause of diphtheria in man? Might not the character of the bacilli be determined by the nature of the soil in which they grow? At present writing it seems that there are three possibilities: 1. Clinical phenomenon of diphtheria without Klebs-Loeffler; 2. Clinical phenomenon of diphtheria with Klebs-Loeffler; 3. Klebs-Loeffler without any clinical phenomenon of diphtheria.[114]

Chappell was one of the first physicians in the city to raise these questions about the bacteriological diagnosis. His questions were difficult to address because the issues involved were not yet well understood. He recognized that if virulence tests had to be conducted to confirm a diagnosis of diphtheria, then one of the key selling points of bacteriological diagnosis—its rapidity—would be jeopardized. As Park had pointed out in his careful studies in 1892 and 1893, many types of bacilli were often found in the throat, and the nature of the so-called pseudodiphtheria bacilli was unknown. Some researchers thought they were typical Klebs-Loeffler bacilli that had lost their virulence. Park advocated a cautious approach: "The early diagnosis from cultures merely forces us to do what other considerations would cause us to do voluntarily, namely, to regard as true diphtheria not only those due to undoubted Loeffler bacilli, but also those unusual cases due to a bacillus which is either closely allied to, or a modification of, the Loeffler bacillus."[115]

Park, assisted by Alfred Beebe, responded to the questions raised by Chappell and others in an extended bacteriological report on cases of suspected diphtheria examined between May 1893 and May 1894. Of the 5,611 cases, 60 percent were found to be true diphtheria and 40 percent false. The cases were about evenly divided among males and females, who ranged in age from three weeks to seventy years. Mortality was highest in the first two years of life, then steadily diminished until adult life, when it slowly increased again. Eighty percent of reported cases of membranous croup were found to be diphtheria.[116] (This finding resulted in an amendment of the sanitary code in June 1894 declaring all cases of croup to be reported as laryngeal diphtheria.)

The report detailed the history of the bacteriology of diphtheria beginning with the work of Klebs and Loeffler and confirmed the Klebs-Loeffler bacillus as the undoubted etiological agent in diphtheria. It described in detail how cultures were made and analyzed, and it addressed a number of questions in a set of original investigations. Among the questions addressed were: How much reliance could be placed on bacteriological diagnosis? If bacilli were found in cultures that possessed the shape, size, and staining characteristic of the diphtheria bacillus, could they, without further animal or cultural experiments, be considered virulent diphtheria bacilli? How long did virulent diphtheria remain in the throat? What was the relation between pseudo and nonvirulent bacilli and the true virulent bacilli? How communicable was pseudodiphtheria?

Their results supported the following conclusions: The examination of cultures did provide a reliable way of determining the presence or absence of the diphtheria bacillus. Pseudodiphtheria was defined as all inflammations of the mucous membrane that simulated true diphtheria but were caused by other microorganisms. It was not as fatal or as easily transmitted as true diphtheria. With regard to the question of virulence, Park and Beebe cited the work of other well-known researchers, which they confirmed by their own experiments. "All bacilli found in throat inflammations suspected to be diphtheria, which possess the morphological and cultural characteristics of the Loeffler bacilli, must be regarded as virulent," they concluded, "unless animal inoculations prove otherwise."[117] In other words, to answer Chappell's question, they thought that in most cases virulence tests were not necessary to make an accurate diagnosis. This report provided the most detailed evidence about bacteriological diagnosis of diphtheria. Bacteriological diagnosis was reconfirmed as the only method that could reliably be used to identify diphtheria.

Conclusion

While cholera had brought about the establishment of New York City's bacteriological laboratory, it was diphtheria that established its importance to public health and medical practice. In two short years, from 1892 to the summer of 1894, diphtheria had been transformed. Because of the program initiated by the New York City Department of Health to provide bacteriological diagnosis of diphtheria, physicians in the city were in a unique position to observe this transformation. Physicians wanted some way to distinguish diphtheria cleanly and sharply from other diseases of the throat. No anatomical distinction could be used definitively to identify diphtheria. The New York City diphtheria diagnostic program, through its laboratory and the wide dissemination of Park's "culture kit," provided the means to distinguish diphtheria from other diseases. Yet, in doing so, it introduced new problems.

The bacteriological evidence indicated that there were two forms of the disease, "true" and "false" diphtheria. This fragmentation of diphtheria into two forms was only possible when the disease was defined in terms of the presence or absence of the Klebs-Loeffler bacillus. For Park, this fragmentation of diphtheria was necessary, based on the evidence of laboratory studies and his clinical observations. In the laboratory, only the Klebs-Loeffler bacillus produced diphtheria. Klebs-Loeffler bacilli, streptococci, and other bacteria produced something that clinically looked like diphtheria among cases in the city hospital wards. Park concluded that there was only one "true" diphtheria and many varieties of "false" diphtheria. The earlier debate over whether membranous croup or follicular tonsillitis were "true" diphtheria had simply shifted from the bedside to the laboratory.

The laboratory could define what "true" diphtheria was by fragmenting the notion of disease itself. First, the Klebs-Loeffler bacilli seemed to take different forms: one that was virulent and one that was not virulent. "True" diphtheria was caused by virulent bacilli. Second, it fragmented the notion of disease because individuals could harbor the bacteria with no adverse symptoms. It thus broke the notion that disease was concurrent with its symptoms. Park and Beebe found virulent bacilli in about one percent of the healthy throats in New York City. Yet these people, they asserted, did not have diphtheria; "true" diphtheria could not simply be defined as the presence of the Klebs-Loeffler bacillus but had to be accompanied by clinically observable symptoms. Yet the presence of the bacillus marked the individual as a source of danger to others. Only the absence of the bacillus could remove that stigma.

Ironically, this bacteriological fragmentation of the disease did not carry over into the management of diphtheria in actual practice. Once an individual was identified as having the bacilli, whether symptomatic or not, he or she fell under the purview of all the regulations that constrained and stigmatized those with the disease. Most physicians wanted to avoid having their patients isolated simply because of a bacteriological report. They would insist that the disease of diphtheria was more than simply the presence of the bacillus.

The implementation of the diphtheria diagnostic program also transformed medical and public health practices in the city. The diagnosis of diphtheria was no longer under the sole control of the physician. Once bacteriological examinations were employed, the means for clear definition of the disease was established. The removal of children to hospitals and the barring of them from schools was standardized. Medical inspectors were to rely solely on bacteriological reports and enforce all the health department regulations. Few spoke out against the aggressive enforcement of these regulations in the tenement districts. For people living in these districts, the presence of diphtheria now resulted in greater intervention of public health authorities into their lives.

Implementation of the diphtheria diagnostic program also exposed the tension between the bacteriological conception of the disease and epidemiological data. As part of the program's implementation, cases of diphtheria were mapped in each region of the city. Patterns emerged that could not be fully explained by bacteriology. The epidemiological information gained as a result of the bacteriological program confirmed the view that some blocks in the tenement districts were "diphtheria nests." Within six months, the health department maps showed regions in the city in which nearly every block occupied by tenement houses contained one or more cases of diphtheria; in some instances fifteen to twenty-five cases had occurred. Local epidemics could be charted, some originating in a neighborhood, others centered around a particular school. As health inspectors tried to pin down the connections between each case of diphtheria, they found that ordinarily the susceptibility to diphtheria varied widely in the city and that bacteriological examinations could not help them locate the source of outbreaks with any certainty. Epidemiological information gave support to what had been long known about diphtheria; that is, it varied widely in most populations. The epidemiological data in some ways weakened the bacteriological evidence, which could not fully explain this difference in susceptibility.

It had been hoped that bacteriology, which had provided the crucial information about the cause of diphtheria, could also lead to its control. The emerging epidemiological information gleaned from the city's mapping effort showed that the prevalence of diphtheria in the city continued to increase despite the bacteriological program and its attendant practices aimed at controlling the disease. The practices used to control the disease — removal to hospitals, isolation, and disinfection — were applied more widely. Yet the sum of these efforts proved insufficient. Park and Beebe concluded:

> With the results of these investigations before us, we can appreciate the difficulty of exterminating diphtheria from a city like New York. On the one hand, we have cases of diphtheria scattered all through the city, many of which are so mild as to be unrecognized, and, on the other hand, we have the crowded tenements with their ignorant and shifting population, where proper isolation of the patient from other members of the family, or of the family from other inmates of the building, is usually impossible unless harsher measures are adopted than are now customary. With stricter isolation of patients and intelligent and systematic supervision of the schools and tenements, we can certainly reduce the number of cases of diphtheria in the city, but the total extermination of the disease under the existing conditions of life here does not seem probable unless we can acquire new means to combat the disease.[118]

Ultimately, even those most committed to the power of bacteriological diagnosis and the promise it held to control the disease had to concede defeat. They would have to be content with the reduction of diphtheria. The goal of absolute control was not possible with existing methods.

3: THE SELLING OF THE ANTITOXIN

No aspect of the history of diphtheria garnered as much public notice and attention as the development of diphtheria antitoxin. The announcement of the successful treatment of children with this remedy at the International Conference of Hygiene and Demography in the fall of 1894 was a significant event with important implications for medical research and for medical and public health practice. The antitoxin was one of the first, and surely the most visible, products of bacteriological research and scientific medicine during this period. It was also a news event, with newspapers around the world heralding the new drug.

While historical accounts of the introduction of antitoxin as the orthodox therapy in the treatment of diphtheria have pointed to the role played by newspapers — most notably in organizing subscription campaigns to raise funds for the production of the remedy — few have examined the critical role these same news reports played in gaining the acceptance of antitoxin by physicians and the public. Most writers have concentrated on either the news reports or the medical responses to antitoxin. This account differs from previous analyses by bringing the two together.[1]

Diphtheria antitoxin was introduced in the United States in New York City in the fall of 1894 and the winter of 1895. Many sources point to the public campaign by the *New York Herald* to raise funds for the production of antitoxin and its free distribution to the poor as a sign of the public's support of this new remedy.[2] A review of the media accounts reveals a more complex story reflecting more than the preexistence of public support. The antitoxin campaign was orchestrated by supporters of antitoxin to *create* support for the new drug among physicians, city politicians, and the public. In this chapter I extend previous discussions by focusing on the creation of support for the new drug. The process of medical innovation exemplified in the case of diphtheria antitoxin, I suggest, must be seen in terms of the social interests that linked parents, physicians, public health authorities, and newspapers. The public subscription campaigns should then be viewed as one of the sites where these interests were revealed. More important, these campaigns highlight the ways in which the successful control of diphtheria involved forms of persuasion that shaped the public, political, and medical climate in which the interventions were introduced.[3]

Public subscription campaigns to support antitoxin production and distribution were not unique to New York City. Similar campaigns were held in the same period in both France and Germany. These campaigns were instituted to finance the new therapy, to improve and boost the image of medical research, and to mobilize municipal and state support for medical research in infectious diseases.[4] The public subscription campaign in New York City turned the new therapy into a cause that disparate and contentious social groups could support by linking the product of medical research to the unimpeachable humanitarian effort of saving children's lives.

An examination of the newspaper coverage of the introduction of antitoxin in New York City reveals the context in which the new remedy was presented to New Yorkers and to the rest of the country. The New York City Health Department attempted to obtain public funds for the production of antitoxin in September 1894, but the decision on the appropriation was delayed until December because the scientific merit of the project was deemed questionable by the politicians controlling the city's funds. However, the health department proceeded with the project with limited funding from private sources, and the *Herald*'s popular subscription campaign kept the issue before the public. The relationship between the newspapers and the health department was a complex one. Historians and the biographers of Biggs, Prudden, and Park have tended to elevate, post hoc, the role of the newspaper campaigns as evidence of public support for antitoxin without examining the role of these men in orchestrating them.

In the end, the campaigns were successful in raising funds and public consciousness about antitoxin. Yet in doing so, they also sensationalized and overstated the efficacy and achievements of the new drug. Newspaper reports downplayed side effects in the promotion of the antitoxin's curative power. The appeal was couched in the value-neutral terms of laboratory research, which elevated the status of this research and of the experts involved in antitoxin production at the health department. It explicitly provided Hermann Biggs with a public platform from which to promote bacteriological research and to defuse potential critics of his new disease-control program located in the health department. The rhetoric employed by Biggs in the subscription campaign promised that the success that bacteriological research had brought to diphtheria would soon be extended to other infectious diseases as well. Though these claims proved to be excessive, they nonetheless helped to establish diphtheria antitoxin as the cultural symbol of the power of medical research and to elevate the public role of the medical researcher to that of the "conqueror" of disease.

Antitoxin Unveiled: The Eighth International Congress of Hygiene and Demography

Research on diphtheria flourished between 1888 and 1894. A large number of investigators worked on questions related to all aspects of the biology of the diphtheria bacillus, its incidence in cases of diphtheria and other related conditions in the upper respiratory passages, and the production of the disease in animals. Indisputably, the most important work was done by Émile Roux and Alexandre Yersin, beginning in 1888 at the Pasteur Institute, where they conducted experiments that provided conclusive evidence that the diphtheria bacilli produced a soluble toxin.[5] For many of the researchers who followed them, the most interesting line of inquiry lay in uncovering the nature of the toxin that the bacillus produced.

In the winter of 1890 the German researchers Emil Behring and Carl Fraenkel laid out an entirely new approach to this question. They sought to create an artificial immunity against the poison produced by the diphtheria bacillus. The culmination of this work radically changed the study of diphtheria and produced a contentious debate over the question of priority in this work. On December 3, 1894, Fraenkel published a monograph describing the inoculation of guinea pigs with attenuated cultures of diphtheria.[6] He demonstrated that two weeks after the injection, these animals withstood injections of virulent living diphtheria cultures without developing symptoms of the disease.

On December 4 Behring and his colleague Shibasaburo Kitasato published their famous paper on the question of immunity in tetanus and diphtheria.[7] Though this paper dealt primarily with tetanus immunity, its implications for diphtheria were apparent. The authors found through their studies of immunity in animals that the immunity of rabbits and mice that have been immunized "rests on the ability of the cell-free fluids of the blood to render harmless the toxic substances which the tetanus bacilli produce." This capacity was so durable that it still persisted when the immune serum was transferred to other animals. Using this immune serum, Behring and Kitasato were able to protect mice that had received three hundred minimally lethal doses of the tetanus poison. Normal serum had no such properties. In a footnote to this paper, Behring and Kitasato coined the name of the new serum: "One may designate these actions against bacterial poisons as 'antitoxic' or 'antifermentative' in contrast to 'antiseptic' and 'disinfectant,' both of which refer to antagonistic action against living infectious material." Thus the new immune serum was named "antitoxin."

A week later Behring published a paper on immunization against diphtheria that confirmed all the observations made about tetanus. In this paper he illuminated five different methods by which immunity against diphtheria could be obtained in animals.[8] The possibility of serum therapy on humans opened up by Behring and Kitasato's work was immediately pursued by German and French scientists. The story of Behring's successfully treating the first child with antitoxin in a Berlin clinic on Christmas night in 1891 is often repeated, though never verified.[9] The story was widely reported, however, and it certainly contributed to the image of the serum as the savior of children.

Behring and his colleagues had increasing success with producing antitoxin between 1891 and 1894. Though a suitable large animal host had not been found, trials on human patients continued in 1892 and 1893. Hermann Kossel, one of Koch's assistants working in the Charite Hospital in Berlin reported the results of 233 cases where he demonstrated that survival rates were higher when the antitoxin was administered early in a case.[10]

By early 1894 Emile Roux and his colleague Louis Martin in Paris had succeeded in producing a potent antitoxin using immune serum from horses. The horses were inoculated with toxin over a two-month period. The antitoxin's potency was demonstrated by the fact that it even "melted away" the false membrane.[11]

In their second paper (with Chaillou), Roux and Martin reported on the treatment of three hundred children with the serum. Beginning in February 1894 and July 1894, they treated every case of diphtheria at the Hopital des Enfants Malades in Paris, with antitoxin. These months spanned the period that typically included both the highest and lowest incidence of the disease in Paris. Four hundred forty-eight cases were treated with antitoxin; 339 recovered and 109 died, giving a case fatality rate of 24.5 percent. In previous years the rate had ranged from 45 percent to 55 percent at this same hospital. The results of this study were compared with those from the Hopital Trousseau, where no antitoxin was used in diphtheria cases. There, 520 cases of diphtheria were admitted and 316 of these patients died, giving a case fatality rate of 60 percent. Roux also found a significant reduction in the death rate with antitoxin treatment in cases complicated by severe streptococcal infections. In conclusion, he noted that there was hope for even greater reduction in the death rate with the new treatment.[12]

Roux reported on this work at the Eighth International Congress of Hygiene and Demography held in Budapest on September 1–9, 1894. By all accounts the response to his talk was instant acclaim and international attention. Victor C. Vaughan, one of the American physicians present, later wrote, "Hats were thrown to the ceiling, grave scientific men arose to their feet and shouted their applause in all the languages of the civilized world. I had never seen and have never seen since such an ovation displayed by an audience of scientific men."[13]

With Roux's report in Budapest, diphtheria antitoxin became international news. The French press immediately took up the story, proclaiming Roux the discoverer of the lifesaving serum despite Roux's judicious attempts to give credit to Behring and the other German researchers. This wholesale adoption of the new remedy by the French was viewed by the British as excessive and "theatrical."[14] But despite this criticism, the media attention given to antitoxin fueled the demand for the new remedy around the world.

Antitoxin for New York City

In the late summer of 1894 when Hermann Biggs made his annual visit to
Europe, he found that everyone in medical research circles in Germany and
France was discussing the extremely favorable results in the treatment of
diphtheria with the new antitoxin remedy. During June he visited the Institute for Infectious Diseases in Berlin, where he observed the production of
antitoxin firsthand. At that time about four hundred cases of diphtheria had
been treated with the serum at the institute under the direction of Behring
and Paul Ehrlich. To Biggs the results were astonishing. He knew that Roux,
working in Paris, had also successfully treated a number of cases. Biggs was
convinced by what he observed: "I found then that there was no shadow of
doubt in the mind of any of these observers as to the specific value of antitoxin in the treatment of diphtheria."[15] Biggs cabled William Park in New
York, instructing him to begin the process of cultivating diphtheria bacilli
for the production of toxin at once, so that upon his return they could begin
immunizing animals for the production of antitoxin.

Biggs also cabled Commissioner of Health Cyrus Edson about the new
remedy and about his desire to begin production of antitoxin in New York
City. Edson then presented Biggs's proposal to both the Board of Health and
the Board of Estimate, requesting an appropriation of $27,000 to begin the
production of the new serum.

On August 27 Edson called a press conference to make public the details
of his request for funds. The New York Times reported in a front-page article
entitled "Sure Cure for Diphtheria," that "diphtheria will have lost all its terrors if information just issued by the New York City Board of Health shall be
proved reliable. Instead of being among the most dreaded of diseases with a
percentage of fatal results almost unequaled it will be reduced to one of the
most ordinary ailments . . . in all cases where the medical treatment and the
proper remedy can be obtained."[16]

Diphtheria had never before been so prominently featured in the New
York press. "Epidemics of diphtheria need no longer be feared," the article
continued, claiming that, in the past, when a case of diphtheria developed
in a household, "it has generally attacked every member of the family. In the
tenements, where it is most often met, it has frequently swept through every
floor, despite the most rigid precautions, and children especially have been
carried off by dozens." Since diphtheria had been so rarely mentioned in the

press before, the exaggeration of its presence in the city is striking. Readers were offered no explanation why a disease with "a percentage of fatal results unequaled" by any other disease had not received more attention. In addition, little of this kind of hyperbole had been evident in the previous year when the bacteriological diagnostic program had been instituted by the health department. In this first announcement of the antitoxin, the representation of diphtheria was changed. It was no longer the familiar dread disease. It was now the disease that could be treated by the "most important discovery made in medicine," and as such was made visible in an unprecedented way. Diphtheria was news now because it could be cured by antitoxin. In particular, as the headline indicated, it could be cured by the "sure cure."

Reports of cures for diphtheria had appeared regularly, especially in the penny press, and to date little had been made of the reputed claims of these medicines. It was important for Edson and Biggs to emphasize that antitoxin was not like other cures. The *Times* erroneously reported that the serum was discovered in Dr. Koch's lab, where the failed tuberculin was also produced. Edson made it clear that the new cure for diphtheria was produced by Koch's associates, but under tightly monitored conditions. "Because of the fiasco of the famous 'lymph' treatment for consumption," Edson said, "which was made public before the experiments had been concluded, the work on the diphtheria investigations was closely guarded." Equally important to the antitoxin's curative power was its value as a preventive treatment as well. "Hereinafter," Edson noted, "no fear at all need be entertained of the disease spreading, as the persons exposed to it can be rendered absolutely immune by inoculation with the anti-toxine."[17]

Thus on August 27 the health department laid out its plan for the production of the antitoxin, strategically asserting its knowledge and authority. The unique expertise of the bacteriologists in the New York City Health Department and the necessity of health department control over the production and distribution of antitoxin were the key points in the argument. Edson stressed the importance of the discovery of the serum, the miraculous results that could be obtained in both cure and prevention of diphtheria by its use, and why it was important that the New York City Health Department begin producing the serum. He informed the press that Biggs had been sent to Europe by the health department specifically to investigate the new remedy. Though this was not the case, it served to create the impression that only the health department had the expertise to produce the serum. Second, Edson

argued that because of the expense associated with the production of the serum, only a municipal or state sanitary authority could produce it in sufficient quantities to meet the needs of the poor.

In an editorial following the press conference, the *New York Times* voiced
its support for the health department proposal—not on the merits of the
remedy, which of course it was in no position to evaluate, but because Edson
had proven himself to be a sensible and cautious physician and because, if
his estimate that the remedy would save the lives of 1,500 children was correct and the remedy was infallible as he claimed, then the city should spare
no expense in procuring it.[18]

The Board of Estimate refused Edson's request for immediate funding but
left open the possibility that the project might be funded later. The board
indicated its willingness to grant the funds if "it was shown that the scientific
work was necessary."[19] The board was to meet again on December 20 to make
the final appropriations for the coming year.

It is important to note the content of Biggs's proposal that Edson presented
to the Board of Estimate. Biggs requested funding not just for the production
of antitoxin but also for "bacteriological studies in diphtheria and tuberculosis, and other lines of research allied to these studies, for instance tetanus and
typhoid fever, and which are likely to lead to good results."[20] It is clear by this
proposal that the city had not really made a commitment to fund the kind of
bacteriological work that Biggs envisioned his division engaging in.

Biggs and his colleagues in the health department were shrewd enough
to recognize that politicians controlling the city's funds needed more than a
simple assertion of the "scientific value" of the work he wanted them to support, but he was not prepared to wait four months to begin producing antitoxin. Two sources claim that he and Prudden purchased the first horses and
equipment to begin the work with their own money.[21] By mid-September a
number of horses were being prepared, but Biggs knew that these few horses
could not produce enough antitoxin to serve the city's needs. Clearly, an alternative source of funding was needed to support both the production of
antitoxin and the ongoing projects of the bacteriological laboratory.

Making Diphtheria "Visible"

While the attention of the medical and lay community was fixed on the developments with diphtheria antitoxin during the next months, the health de-

partment also instituted a new practice that dramatically increased the visibility of diphtheria throughout the city. On September 28, 1894, the Board of Health implemented the policy suggested by Hermann Biggs and Alvah Doty, chief inspector of contagious diseases, of placarding tenement houses where cases of diphtheria, scarlet fever, or measles were found.[22]

This new policy was needed, it was argued, because of the failure when inspectors from the health department simply visited families where cases of these diseases had been reported and informed them of the necessary precautions. Not only were other families in the tenements unaware of the presence of contagious disease, but strangers and visitors to these houses were also oblivious. Verbal notification that a case of a contagious disease was present in the building was usually ignored or forgotten, and the degree to which inspectors actually followed through on such notification was also unclear.

The new ordinance was clearly directed toward the concerns and protection of the middle- and upper-class population rather than to the poor. The Board of Health circular noted that previous methods had failed to protect ladies from contagion in their search for servants. Nor did former methods protect families from servants who brought infection from houses where there were cases of diphtheria. The circular further warned that "very commonly, washing or various kinds of sewing is secretly done by other members of the family in apartments where such cases are ill, and the garments thus infected on the premises are later returned to the owners." Small businesses located in such tenements were also cause for concern: "In small shops, business is sometimes carried on, and in one of several instances recently, a number of cases of diphtheria were directly traced to an infected candy store."[23]

Since the health department's bacteriological diagnostic program had demonstrated that diphtheria patients often appeared to be well before they were free of the diphtheria bacilli, placarding also made it possible for building residents to continue enforcing isolation of those cases until the inspectors removed the placards. The health department ordinance stressed the moral influence placards would have on the "inmates of the apartments, the inmates of the house, and strangers and visitors" by encouraging them to practice isolation because of the placard's visibility.

Publicly declaring a house to have a case of contagious disease was problematic, as Biggs well knew. To date no stigma had been attached to the presence of diphtheria, scarlet fever, or measles in a household. Yet the use of placards evoked negative attitudes. Biggs acknowledged the approbation directed toward households placarded for the presence of tuberculosis, yet he was not deterred by such sentiment:

> For some months in certain classes of tuberculosis the system of placarding apartments has been authorized and employed by the Health Department, and has proven very satisfactory in the attainment of the object desired. The only objection apparently to be urged against this measure is that the inmates of the apartment may object to the publicity thus entailed. This, however, is exactly the object which the measure is justly and properly designed to subserve, and is, in our opinion, the strongest argument in favor of its adoption.[24]

Biggs wanted to make sure that everyone could see the prevalence of contagious diseases in the city. Germs and the danger they posed could not be seen by the ordinary person. The epidemiological data that had been accumulated as a result of the diphtheria diagnostic program had made the prevalence of diphtheria, and the danger from convalescent cases in particular, more visible to the experts with access to that knowledge. Yet these experts — bacteriologists and physicians serving as inspectors — could not hope to control disease alone. Once disease could be seen more easily by the public, and the source of danger marked, the public could more easily recognize the need for following the health department's dicta. The stated goal of the new practice was to have those who were well enforce isolation on those the health department declared to be sick. The well could now more easily avoid the sick. The report was approved by the Board of Health on September 28, and white cards for diphtheria, red for scarlet fever, and blue for measles soon appeared on apartments and buildings around the city.

There is little evidence to suggest that the institution of the placard system by the Board of Health at this particular time was an intentional part of the strategy to garner public support for antitoxin. Yet the placarding must have served the project indirectly. As newspaper coverage of the new serum increased, New Yorkers could also for the first time see the evidence that diphtheria was rampant in their midst. With an average of two to three hundred new cases of the disease occurring every week, and the high prevalence in certain districts, placarding made the problem of diphtheria more evident than it had ever been to the average person.

The Newspaper Campaigns

It is generally assumed that the popular subscription campaign launched by the *New York Herald*, which began on December 10, 1894, contributed most significantly to the public acceptance of diphtheria antitoxin in New York City. Few have noted that the *New York Times* ran a series of articles from September 1894 through January 1895 about antitoxin as well.[25]

The articles in both newspapers clearly supported diphtheria antitoxin and the project of producing the serum in New York City under the auspices of the health department in Biggs's Division of Pathology, Bacteriology and Disinfection. The style in which each paper's arguments were presented is strikingly different, as were the audiences they were intended to influence. Yet both papers were expressly engaged in the project of both educating the public about the value of antitoxin and convincing them that the health department's projects had to be supported with public funds. A close reading of these articles shows that these newspapers were well aware of the need to generate public support for antitoxin before the Board of Estimate's December 20 meeting.

Prior to the bacteriological diagnostic program of the Board of Health, there was little public notice of diphtheria. There had never been a public campaign to draw attention to the disease in the city. Since little could be done, there appeared to be no compelling reason to mount such an effort. The visibility of diphtheria as a result of the bacteriological diagnostic program in the city and the announcement of the new remedy gave a new social dimension to the disease. The newspaper coverage reveals a great deal about public perceptions both of diphtheria and of science. Scientific research, it was argued, had robbed diphtheria of its terrors and would bring about the control of other diseases as well. The Board of Estimate had asked for proof of the scientific value of the work that Biggs wanted funded. The newspapers represented to the public how the scientific value of the new cure could be ascertained.

In early September 1894, the *New York Times* published a long, unsigned editorial describing the antitoxin. It carefully explained the remedy and how it was produced in horses. Since there was no antitoxin available in the city, the article explained that it had been used extensively in European cities and claimed, "Everyone who got it on the first day was saved; . . . If cases can be treated in the first thirty-six hours then . . . the mortality may be reduced to zero."[26] The writer was careful not to make too many exaggerated claims about the remedy, noting that "the effect of the inoculations is not permanent." The article closed with an appeal to the city to provide the $30,000 needed to produce antitoxin in New York City in order to save as many as one thousand lives. The appeal was capped off with a vote of confidence for Edson, Biggs, Park, and Beebe, who were to be trusted in this work because of the international reputation they had garnered as the architects of the successful diphtheria diagnostic program.

Later in September, in another article on the editorial page, the *New York Times* printed statistics on the use of antitoxin in Berlin and Paris. The French data, which are from the report Émile Roux presented at the International Congress on Hygiene and Demography in Budapest, showed that the mortality rate was 24.5 percent at Children's Hospital using antitoxin, versus a 60 percent mortality rate in another hospital where antitoxin was not used. The numerous reprints of statistical reports were neither located in context nor discussed in light of clinical knowledge or experience with diphtheria. They were directed at a lay public, who were led to see proof of the serum's usefulness in the simple comparisons.

During October the *Times* printed a number of articles on the cost of the antitoxin and how it was being distributed in Europe. One article reported that some hospitals were providing antitoxin free to patients since its cost, ranging from $7 to $20 per case, was prohibitive.[27] By mid-October Dr. J. J. Kinyoun was sent from the Marine Hospital Bureau in Washington, D.C., to evaluate the new remedy. In his report from Paris to U. S. Surgeon General Wyman he wrote, "Roux has been too modest, the discovery is one of the greatest in medicine, and has passed through the experimental stage and laid a foundation for a new system of preventive medicine."[28] The *Times* was quite enthusiastic about Kinyoun's report and highlighted his planned travel to Germany to learn more about antitoxin in order to begin production in Washington upon his return. The Academy of Medicine in Paris reported favorably on the serum treatment of diphtheria, universally known in Paris as the Roux treatment, a few days later. Several short articles reported on the public subscription campaign initiated by *Le Figaro*, which by the end of October had raised $50,000 for the production of antitoxin in France. Government funding to provide distribution of the remedy at public expense was also pending at this point in France.

The theme of articles in the *Times* shifted in November from simply emphasizing the value of the remedy to discussions of public funding of antitoxin production. The value of antitoxin is assumed to be proven in these articles by data showing reductions in mortality rates and by affirmative authorities in France and Germany, summed up by Kinyoun's report asserting that "the experimental phase is over."

Yet the *Times*, ever mindful of the need to retain some sort of objective stance, also reprinted a long article from the *London Times* on November 9, suggesting for the first time that some skepticism about the new remedy existed. The *London Times* article presented a less-than-enthusiastic response to the events happening on the Continent. The article first acknowledged the excitement generated by the results of the use of antitoxin as evidenced by the many public subscription campaigns and the considerable sums of money being raised to increase production and distribution of the new remedy across the Continent. Yet it urged restraint in the approval of the new remedy. The writer delineated the facts about how the serum was discovered and tested. He noted that the bulk of the scientific credit went to Koch's school of bacteriology, rather than to the French, despite the "zeal of French journalists to promote Roux." The article went on to present a fairly straightforward discussion of the research of Loeffler, Behring, and Kitasato and the results obtained by Roux at the Children's Hospital in Paris early in 1894. Though the writer was struck by the unanimity of positive opinion on the results of the new treatment, he wanted to see further tests, involving many hundred cases from different sources. He did not agree with the *Times*'s assertion that the experimental phase was over.[29]

The chief criticism that the *London Times* writer noted concerning the antitoxin results was the inaccuracy of comparing diphtheria mortality rates in the years prior to the use of antitoxin, because diphtheria varied so greatly in virulence from year to year. Thus the apparent improvement due to the serum reported in Paris may have been merely a sign of the mildness of the epidemic prevailing there during the time of Roux's test. The writer concluded that, given such considerations, it was impossible to measure the "precise, or even approximate, efficacy" of the new treatment from information available: "When Dr. Behring states confidently that 'the mortality may be lowered to 5% and diphtheria reduced to a comparatively dangerless disease,' he seems to be going beyond the evidence."[30]

Despite the cautious tone of the article, the *London Times* was not suggesting that antitoxin was of no value or denying that lives had been saved. Yet caution was needed precisely because expectations of the new remedy were being pushed so high:

It is premature and unwise to teach people that a sovereign remedy has been found
which will rob diphtheria of its terrors. Further experience may considerably modify
the hopes now entertained, and we ought to be prepared for some disappointment;
. . . the new treatment is still on trial, but a strong prima facie case has been made in
its favor.[31]

The British were urged to get involved in the production of antitoxin so
that tests under the supervision of British physicians could confirm the results obtained by workers on the Continent. In apparent agreement with the
position taken by the *New York Times*, it called for the British government
to supervise the production of antitoxin and argued that government control was necessary for a number of reasons. First, applications were already
pouring into the British Institute of Preventive Medicine, and the demand
exceeded the available supply. Second, because of the labor and expense involved, the production of antitoxin could only be carried out by the government. And finally, there was a need for government regulation to protect the
public from the danger of those that some might try to foist improperly prepared substances on the public.[32]

The *New York Times* made no editorial comment on the British perspective, though the headline lead-in to the article downplayed the skepticism of
the British and emphasized their support of antitoxin. The caution expressed by the British, and the points raised about the evaluation of antitoxin, did not influence any of the subsequent articles. They continued
through the month of November to report on trials of the new remedy. Toward the end of November, Dr. A. Campbell White purportedly used antitoxin for the first time in New York City on twenty cases at the Willard Parker
Hospital. Antitoxin was the only treatment given to the children with severe
cases of diphtheria. Only five of the children died, whereas typically less
than half would have been expected to survive. Campbell commented that
he believed that antitoxin, given early in the disease and in sufficient quantity, prevented death by "the absorption of the toxine of diphtheria." The
Times lauded Campbell's results, calling them the "first important and conclusive contribution from America. . . . We are glad it has been furnished by
New York."[33]

A little over a week later the first batch of diphtheria antitoxin imported
from Germany arrived in New York City. The firm of Lehn and Fink announced that seven vials were available for sale at a "nominal" price. The
firm claimed they would only sell to physicians who could prove they had

actual urgent cases of diphtheria to treat. Though it is not clear from the news report whether they followed through on this restriction, there was obviously a need for more than seven doses of the serum in the city. Dr. Seneca Powell, president of the New York County Medical Society, and one of the purchasers of the serum, remarked on the scarcity of the remedy: "I believe I should favor the application of the city's money for the manufacture of antitoxine here, and the Board of Health is the proper department to undertake it."[34]

Powell and a few other notable physicians in the city made several public statements acknowledging that there was no available supply of diphtheria antitoxin in New York City in the fall of 1894. George Shrady, editor of the *Medical Record,* wrote that by November the only antitoxin that had been used in the United States had come from abroad and was then in such limited quantity as to be "practically useless for the purposes of extended experimentation."[35]

Historians and the biographers of Biggs and Park have consistently claimed that the health department produced the first diphtheria antitoxin in New York City and the United States.[36] While it is indeed true that the antitoxin produced by Park and his associates was the first produced by a municipal health department, another source for diphtheria antitoxin in the fall of 1894 was the New York Pasteur Institute.[37] Paul Gibier, the director of the Pasteur Institute also traveled to Europe in the summer of 1894, where he too learned Roux's method of antitoxin production. Gibier claimed to have made a few doses of the serum as early as October, and by December 1894 he was producing a "considerable quantity" of it.[38] For reasons that are not exactly clear, there is virtually no mention of Gibier's serum by Biggs and Park or in the newspaper accounts about diphtheria in New York City during the fall and winter of 1894.[39] I will return to this subject later when I discuss the *Herald*'s campaign.

In early December the need for experts to oversee the production of antitoxin in the city was buttressed when reported sales of a liquid purported to be antitoxin became public. Just as had been noted in the *London Times,* this was a potentially serious problem. Edson, Biggs, Prudden, and others recognized the dangerous situation brewing in the city and immediately notified the Board of Health. Biggs had purchased some of this product and upon analysis found it to be useless. Vials containing three grams of the liquid were on sale in the city for five dollars, with labels featuring the misspelled names of European physicians.[40] The Board of Health then charged Biggs and Prudden with the responsibility of establishing a system to guarantee the purity of products labeled antitoxin sold in the city.

On December 5 Biggs and Prudden provided the board with an update on the status of antitoxin in the city. There was no adequate supply, they reported, and the supply from Europe was woefully inadequate. Additional money was needed for production of the serum to continue. The report also introduced two new factors. First, Biggs and Prudden noted that the price of antitoxin was costly because of the time required for preparation and the cost of materials, animals, and experts. They did not expect the market price to come down anytime soon. Second, given the appearance of spurious antitoxin in the city, it was important for the Board of Health to intervene in controlling the product as part of its role as guardian of the public health. In the first request for funds in August, Biggs had emphasized the value of antitoxin and the practical results to be gained from the scientific study of disease. He had not discussed the economic aspects of serum production or the need for municipal control of the product. With the attention given to the remedy in the press, and the increasing public demand generated by this media attention, the more traditional functions of the Board of Health came to the fore. The board responded by ordering Biggs and Prudden to submit a plan for determining the strength and purity of all antitoxin on sale in the city.

Historian David Blancher has characterized this report as an ingenious stance on the part of Biggs and Prudden to garner support for the production of antitoxin.[41] Rather than supporting what some might consider speculative scientific work, the production of antitoxin could be seen as an extension of the traditional functions of the Board of Health, namely to provide relief to those sick in the city who most needed it and to protect the public health from unscrupulous vendors. Biggs and Prudden now argued that the only logical way for the Board of Health to fulfill its proper role was to fund a laboratory to produce the antitoxin.

The final meeting of the Board of Estimate was approaching. On December 2, 1894, just prior to the report just discussed, Biggs released a long statement to the press describing the plans for the production of antitoxin in the city by the Board of Health. For the first time, he revealed the work that was already being carried out in the city to the public. "New York will have antitoxine by January," he wrote, "through private rather than municipal public spirit, while I ought to say that so far, little, if any blame for delay attaches to the city government." The health department would control distribution of this "boon to humanity." Cases at the Willard Parker and North Brother Island public hospitals were to be the first to receive the antitoxin. By providing serum at these locations, Biggs also hoped to support the use of bacteriological diagnosis in diphtheria cases: "This no doubt will result in every child or adult having the first assured symptoms of diphtheria being hurried to the Health Department's hospitals. So much the better. They will be certain of proper treatment and the staffs will have the benefit of the clinic."[42] In typical fashion, Biggs neglected to note that the Willard Parker Hospital and the North Brother Island facility were incapable of serving even the current level of diphtheria cases in the city.

Once hospital cases were treated, additional poor of the city would be provided with antitoxin free of charge either in city hospitals or through their own physicians. Any excess antitoxin produced by the health department was to be sold at cost for the benefit of those who could afford to employ private physicians, just as with other vaccines. Biggs then explained plans to make antitoxin available in the city in less than one month's time. Until this point, as articles in the *Times* revealed, the public knew only that Biggs's request for funds to begin production of antitoxin had not been granted. In August, few people could recognize the significance of the city's action. But in December, after months of articles in the *Times* and other New York newspapers, many were familiar with the new remedy and its importance for the treatment of diphtheria. Biggs explained how he had returned from Europe after studying the new remedy and requested funds from the city. As noted earlier, his request was not just for the production of antitoxin but for support of other bacteriological research as well. He carefully noted that the Board of Estimate had not rejected his request outright but had deferred the decision until it could be shown "that the scientific work was necessary."

Meanwhile Biggs had secured private funding to enable immediate production of the serum. He explained: "A citizen, who prefers to remain anonymous, reflected that, as it takes months to produce antitoxine, all this time would be lost, and the funds necessary to start the production of serum were supplied, and we are now in a fair way of having a product of antitoxine that we can control in three or four weeks."[43] Thirteen horses and a number of sheep, goats, and dogs were being stabled at the New York College of Veterinary Surgeons. The horses had been inoculated by Biggs and Park. Though he could not predict how much serum would be available, he could assure New Yorkers that there would be an adequate supply for current cases for a short time.

In closing, Biggs professed his absolute confidence regarding the value of antitoxine and presented data collected in Paris from 1889 through 1894, which he claimed "could be trusted to prove its efficacy." Finally, he urged the public to support the appropriation of city funds to continue the work:

> There can be no question that New York needs an establishment of the kind for which an appropriation of $30,000 was asked to carry on scientific work in all these lines that are likely to lead to practical results in the investigations of contagious diseases. . . . [T]he results obtained—let us select the cultures of diphtheria for an instance—fully justify the establishment of an institution.[44]

Again Biggs had proved his skill at presenting his projects to the public. By waiting until the beginning of December to reveal his plans, he was able to capitalize on the media attention given to antitoxin in the press and frame the health department's production of the serum as expressing concern and commitment to the public welfare. The "anonymous citizen" who allegedly provided money for the horses was undoubtedly Biggs and Prudden, according to their biographers.[45]

On the day after Biggs's report was published, the New York Times ran an editorial in support of his plan. Citing his comments and the national and international attention given to the bacteriological diagnostic program, the editorial lauded the official bacteriologists of the city for being "abreast of the times." The editorial carefully summarized the paper's position on the antitoxin thus far: "We have . . . described fully the nature and origin of this antitoxine and directed attention to the statistical and other proofs of its efficacy."[46] The editor unequivocally supported the plan to have the production of antitoxin under the control of the "competent and trustworthy bacteriologists, who have had an opportunity by association with Prof. Behring or Dr. Aronson in Germany or with Dr. Roux in Paris, to become familiar with approved methods of production and administration." The Times further advocated public funding for this work as had been the case in France and Germany. Throughout the fall of 1894, the paper had consistently echoed the theme that only the views of the appropriately trained experts should be given credence in evaluating the new remedy.

The Times coverage of the new remedy for the treatment of diphtheria had centered around presenting certain scientific facts about the efficacy of antitoxin, thereby fulfilling its expository function for the benefit of an educated public. Reports from eminent scientists abroad showing large reductions in mortality rates were presented for review. This approach assumed the readership was equipped to discern the veracity of the information presented. The medical history of diphtheria presented so cogently in the *London Times* article, for example, is largely missing from the *New York Times* reports. Either too much detail is given, as in articles reprinted from medical journals, or too little detail is provided, as in general statistics presented from Paris hospitals.

Specific discussions of medical debates regarding the distinction between diphtheria and other diseases, differences in the virulence of outbreaks, questions about the bacteriological conception of the disease and the new diagnostic program, and the failure of public health efforts to control the disease were not addressed at all in newspaper reports. In particular, the *Times* articles failed to explicate the complexities of diphtheria that were subsumed under the rubric "dread disease." Instead, the ravages of the disease appear in the margins of the discussion of the antitoxin in the newspapers.

Throughout November, for example, an epidemic of diphtheria with seventy-six cases and twenty-one deaths occurred in Yonkers, resulting in the closing of all public and parochial schools.[47] Two weeks later an outbreak in the village of Catskill in Ulster County, New York, left four of the six children of Ira Snyder dead, as well as Miss Sarah Hasbrouck, a local teacher who had volunteered her services to nurse the sick.[48] These stories were reported but not framed in the context of the antitoxin campaign.

The emphasis in these articles was that the health department, led by Edson and Biggs, represented men whose scientific training held them to a standard that transcended partisan politics. Bacteriologists in the health department were represented as exemplars of the new class of professional public officials worthy of the public trust. The call for public trust in these professional public servants was extended to the antitoxin as well.

The Herald Campaign

The *Herald* campaign began after four months of articles had appeared in the *New York Times* about diphtheria antitoxin.[49] During the fall of 1894, few reports had appeared in the *Herald* about diphtheria or antitoxin. One of the first articles, a reprint from its European edition that described Roux's speech before the International Conference of Hygiene and Demography in Budapest, appeared in October. The report detailed Roux's work developing antitoxin and performing successful trials on children in Paris. In late November the *Herald* printed a second article on Roux's work, this time emphasizing the dramatic drop in the diphtheria death rate with the use of antitoxin. "Two deaths from diphtheria in a week in all Paris!" the writer noted. "Never since the Bureau of Statistics was founded fifteen years ago, has such a figure been reported." Much of this article explained in dramatic style the subscription campaign initiated by *Le Figaro* in Paris. In the French press, Emile Roux was lionized as the heir to Pasteur, who, it was said, had unselfishly given his work to the public. The public response to the appeal for funds to establish a private institute to produce the serum was cast in the most emotional terms: "Mothers and children head the subscription list for the foundation of an antidiphtheritic institute, so eager are they to bring into general use the wonderful serum."[50]

In a self-conscious copying of the French, on Monday, December 10, the *Herald* opened its campaign in support of the production of antitoxin. The announcement read: "Due to the undoubted efficacy of Dr. Roux's diphtheria antitoxine, and the general and unsatisfied demand for such a valuable disease preventive and cure in this country, the *Herald* opens a subscription fund."[51] The goal of the campaign was to raise money to provide antitoxin for the poor, whereas the French campaign was focused on creating a permanent institution to produce antitoxin. (Meanwhile, in Germany, Behring had recently demanded that the state provide the public with the serum.) The *Herald* opened its campaign with a one thousand dollar donation but without mentioning how the money raised was to be used. Instead, the *Herald* created a sense of drama with its assertion that "the details of the plan will be announced later."

The article covered nearly the entire page; it featured a prominent drawing of Roux, smaller sketches of Behring and Cyrus Edson, and a laboratory scene depicting a white-coated figure administering antitoxin to a child. Testimonials from New York City physicians attesting to the efficacy of the antitoxin constituted the main body of the text. These testimonials by physicians offer interesting commentary on the diphtheria antitoxin. Dr. H. Holbrook Curtis, a throat specialist, commented, "It's the universal verdict that the serum is entirely efficacious. . . . [It] has stood the severest tests." Dr. Abraham Jacobi added, "Not a minute should be lost. Every minute may mean a human life." Dr. D. B. St. John Roosa, president of the New York Academy of Medicine urged support of the campaign:

Every decent physician will co-operate with the Herald *in this work, and I feel sure that we have no indecent men in the profession . . . Everyone is talking; everyone is interested, and each physician is anxious to obtain some of the fluid. At present few seem able to get it. It is absolutely impossible to get it even from a dear friend.*[52]

These physicians' comments suggest an enthusiasm and unanimity about the antitoxin within the city's medical community even though none of the physicians quoted had firsthand experience with it. The paper does not hesitate to capitalize on the fact that antitoxin was given more value because of its scarcity. For example, Dr. George F. Shrady was one of the few in the city who had managed to obtain a small quantity of the precious serum.[53] The reporter described the scene at Shrady's office: "He then brought from a room that was securely locked a small vial almost full of a fluid of an amber color and handed it to me. After examining it I handed it back and he immediately locked it up again."[54] Shrady claimed that his was the only antitoxin in the country and that he didn't know where he could get more.

The first article on December 10 provided scant details about what the serum was, how it was produced, or what were its effects on the disease. The story line established and repeated in the *Herald* articles during the campaign turned on three themes: the scarcity of the antitoxin, its efficacy, and the issue of who should oversee and control its production in the city.

The last theme, the control of the production of antitoxin, was discussed by every physician interviewed in this first article. Dr. H. Holbrook Curtis suggested that the pathologists and bacteriologists of the city should be invited to act as an advisory board to the *Herald*. He excluded general practitioners in the city, however, from participation on such a board: "General practitioners to my mind should be free from active connection with the establishment so there would be no room for jealousies of any kind."[55] Dr. Alvah Doty, chief inspector of contagious diseases, worried about conflicts between the *Herald*'s plan and the Board of Health. Doty wanted the *Herald* to cooperate with the Board of Health and the funds placed under Biggs's direction. Dr. Abraham Jacobi suggested that the production be overseen by either the Board of Health or a committee of the New York Academy of Medicine. "There are no factions there," he commented.

Factions evidently existed within the New York medical community, however. The divisions between specialists and general practitioners were embodied in the comments about controlling the antitoxin. The newspaper commentary implied that specialists were trying to assert their power relative to the general practitioners by insisting that only specialists had the expertise to use the new serum properly. Dr. J. Mount-Bleyer, specialist in diphtheritic and throat diseases and visiting surgeon at the New York Throat and Nose Hospital, supported such a view. He argued that specialists should be put in charge of administering antitoxin in hospitals because they were "accustomed to inoculation and the treatment of the throat, and not in the care of general practitioners who would lose too much valuable time."[56] As Charles Rosenberg argues, specialists saw in the new treatment of diphtheria another opportunity to show that they represented technical superiority and the highest quality of care.[57] In contrast, the general practitioner was represented as being behind the times and perhaps not to be trusted with control of the new remedy.

On the second day of the campaign, the self-advertising tone of the newspaper was vividly revealed. The *Herald* featured itself prominently in the story as receiving "universal commendation" for its campaign from physicians, scientists, and newspapers across the country:

> The plan has taken the community by storm. Rarely — perhaps never before — has a
> charitable undertaking so completely engaged the best feelings of the best elements
> of the metropolis. It means the actual salvation of thousands of human lives, especially the lives of the little ones of the poor, who have always been shining marks for
> the dread darts of this most fatal of scourges. It is no exaggeration, but the simple
> truth that many of those lives that are being lost every day might be saved by one
> dose of anti-toxine. This wonderful remedy, the medical faculty admits has stood the
> severest tests, and its efficacy is unquestioned.[58]

The *Herald* created two heroes of the story. In the most dramatic style, it proclaimed antitoxin as the miracle drug with the unanimous stamp of approval from physicians. The *Herald* itself was portrayed as equally heroic for providing this wonder drug, thus making it possible to save the poorest children in the city. Some lesser heroes were allowed to play supporting roles as well. Among them was Biggs, who was prominently featured in the second article. He described his trip to Europe to study the manufacture and use of antitoxin and his desire to produce it in the United States. He presented statistics from Paris that showed decreases in deaths with the use of antitoxin. As one might expect, the data fluctuated somewhat, but he did not explain this, nor did he mention any other problems that would have affected the statistics. The data simply showed a sharp drop in diphtheria deaths with the use of antitoxin. Biggs emphasized the contrast between the Paris data and the situation in New York. Citing only three years, 1891 to 1894, he reported that the number of deaths due to diphtheria had actually risen from 1,361 to 1,970 in New York City. Biggs was obviously not interested in providing a complicated picture for the *Herald's* readers. Few of the technical details that appeared in the *Times* articles were given. Biggs made just two points: antitoxin was harmless and it saved lives. Though he emphasized the positive, he did note the conditions under which antitoxin should be used and expected to give the best results:

> Any person after exposure can be rendered immune to the disease if the symptoms have not already developed. If cases can be treated within the first thirty-six or forty-eight hours . . . the mortality maybe reduced to practically zero. After this time the value of the treatment becomes progressively less. [59]

Biggs did not stress the practical implications of his comments for physicians. The fact that antitoxin was most effective when used early in a case of diphtheria before the onset of the severest symptoms would prove to be an extremely important point later when physicians evaluated antitoxin. The *Herald* did not emphasize that the failure to use antitoxin early would undermine the favorable statistics it was printing showing that antitoxin could reduce the death rate to "practically zero."

Other testimony from physicians in this article continued to build support for the *Herald* and antitoxin, although most of those quoted had not used it. Dr. Urban Hitchcock's comments were typical: "No better work could be done. . . . Surely our public will not be behind the French in their response to such an appeal . . . reports from Paris from such high authorities leave hardly a doubt as to its efficacy." The *Herald* reveled in creating an aura of excitement about the campaign, building anticipation throughout the city for the new remedy. One physician reported patients writing to him daily asking how they could obtain the serum. Another commented, "No doubt of its efficacy. We are all waiting anxiously until we can get it."

To the extent that doubts were voiced from the medical profession, they were quickly dismissed. Dr. George V. Hope, visiting physician at the Metropolitan Throat Hospital, displayed some skepticism about the data in support of antitoxin. Since he had no personal knowledge of its efficacy, he relied on the testimony of those who had used it. Despite his lack of firsthand experience, he asserted that he was not prepared to pass on the value of a remedy that was only of use on the first day of the disease: "Now, what are we to argue from that? Remember that most cases of diphtheria do not reveal

themselves at first, except as a sore throat, and a large percentage do not pro-
claim themselves as such even under the microscope." It is odd that Hope's
comments didn't generate any questions from physicians. Given what was
generally known about diphtheria, most in the medical community would
have seen his comments as being quite reasonable, yet the Herald characterized them as "the plaint of men who resent any incursion into what they are
pleased to regard as their own special preserves, of the vigorous and progressive action which has shown in the past, and will always point the way to
mere formalists." The Herald was not interested in its project being undermined by questions targeting any aspect of the new remedy.

The physicians interviewed also continued to comment on how the Herald should use the money raised in the campaign. The Herald claimed that
the "weight of professional opinion is inclined to favor the appointment of
an advisory commission, composed of physicians, clergymen and laymen,
who will act in conjunction with the appropriate city officials."

It appears from events that followed, however, that some negotiations
were occurring behind the scenes on this issue. On December 13 the paper
announced it would confine itself to collecting money; the administration
of the funds would be left to "well-known and able physicians of the city—
who will act as an advisory committee which will select a subcommittee of
bacteriologists and pathologists and a group of prominent citizens."[60] The
next day the advisory committee was announced. It consisted entirely of physicians, almost all of whom were members of the New York Academy of
Medicine, the most elite of the city's medical organizations. Included in the
list were Edson, Biggs, Park, Bryant, Shrady, and Gibier. This group in turn
selected a five-person subcommittee that was charged to "confer with the
Health authorities and to arrange details for the practical work of cultivating
the now famous diphtheritic remedy."[61] This subcommittee consisted of
Prudden, Park, Biggs, John S. Thatcher, and John Winters Brannan. Three
of these men—Prudden, Biggs, and Park—worked for the health department and were the very ones already involved in producing antitoxin. The
Herald essentially had decided to turn funds from the campaign over to the
health department, though this was not directly acknowledged. The advisory board and its subcommittee were represented in its articles as if they
were separate from the health department. In fact they were not.

Whereas the Times took a strong editorial position that only the "experts"
like Biggs and Prudden should control the production of antitoxin, the Herald had gone to some lengths to make it appear that they favored a more open

process. The previous day's articles suggested that physicians and other leaders in the city were being sought out for advice. The final composition of its advisory board suggest that this was perhaps a ruse. There was never any debate over who should control the production of antitoxin in the city aside from the pathologists and bacteriologists working at the health department.

By December 15 the *Herald* had raised a total of $2,212.50 for the antitoxin fund.[62] Donations received by the paper were often accompanied by letters from parents whose children had died from diphtheria. One donation letter read: "Please accept $1 for your anti-toxine fund from a father who lost a dear boy, five years old, and his golden haired baby girl, two years old, inside of one week from the dreadful diphtheria."[63]

Unlike the *Times*, the *Herald* gave more space to reports of outbreaks of diphtheria throughout the state of New York and the rest of the country. An epidemic in Ashtabula, Ohio, received regular coverage during the month of December. The outbreak there as described in the report below, gave a vivid picture of what happened when diphtheria struck a town with few resources to control it:

> The infected wards are among the best residence portion, and all are well sewered and paved. The other wards are along the river, in a bad sanitary condition but there is no epidemic there. Public meetings have been abandoned for three days; no church services are allowed and schools have been closed in three wards. People of other towns are afraid to come to Ashtabula and shun travelers from here and often refuse express or freight packages. A large force of extra policemen will go on duty to guard houses.[64]

Reports from other towns where diphtheria epidemics occurred served to bring home to the reader the havoc that diphtheria could wreak when uncontrolled. The vivid images in such articles needed no editorial comment to make this point.

By mid-December the *Herald* articles began to feature a different aspect of the antitoxin story each day to maintain readers' interest. On December 15, for example, the main article featured the inoculation of the horses whose blood was used to make the serum. Biggs, Prudden, and Park figured prominently in this story. The article noted that they had prepared eighteen horses in the autumn and courageously "met all expenses themselves despite being turned down for funds by the Board of Estimate." The story is accompanied by a large drawing of a horse being inoculated at the stables of the New York Veterinary College. Perhaps in an attempt to head off any criticism by antivivisectionists, the article notes, "The horses which the bacteriologists of the Health Department have under treatment have a sleek, well contented look. Apparently they have never enjoyed anything so much in their lives as their semi-weekly doses of cultivated poison. They are a plump lot of animals."[65]

The cost of producing antitoxin was the topic of articles during the middle of December. Given that an average of two hundred cases of diphtheria per week occurred in New York, with almost two thousand cases annually, the *Herald* argued that the cost to produce sufficient antitoxin for the city would be very high. The cost of the serum had become an issue in France, Germany, and England as well. Popular subscription campaigns and public appeals had been initiated in Paris, where 500,000 francs were raised. In London Sir Joseph Lister's plea had resulted in a one thousand pound donation by the Goldsmiths Company. Appeals for funds were being made in St. Petersburg, Vienna, Berlin, Milan, Havre, Rouen, and other European cities.[66] The *Herald* framed the issue as a question of "life against dollars" as it urged New Yorkers to increase their contributions to the fund. Several articles described desperate parents searching for physicians who had the precious antitoxin. Stories like the following were common:

> *Inoculations with anti-toxine were made at the Pasteur Institute yesterday on the two children of Mrs. Ann Petersen, of East 104th Street. The patients are a boy and a girl, respectively nine and eleven years old. They are both suffering from malignant diphtheria.*
>
> *The mother, who is a widow, read of the new diphtheria remedy in the* Herald. *The case was urgent, and she was referred to the Pasteur Institute, where she arrived at midnight on Friday, very much distressed over the condition of the little ones. It was too late to treat them at once, and so the applications were deferred until yesterday.*[67]

Though both children received antitoxin, the article did not indicate whether they survived. To the extent that such stories were factual, they suggest that by the end of 1894, the people of New York City indeed believed antitoxin to be an effective treatment for diphtheria and sought out physicians who had it in order to save their children's lives.

This testimony was one of the few acknowledgments of the New York Pasteur Institute's diphtheria antitoxin. Paul Gibier, its director, subsequently published a number of favorable reports of physicians who successfully treated patients with his antitoxin during late December 1894.[68] Though Gibier was selling diphtheria antitoxin at the New York Pasteur Institute in December, there may have been some problems with his product.[69] Biggs and Prudden were most certainly aware of Gibier's product, since he served on the *Herald*'s advisory committee, yet they still maintained that "no trustworthy antitoxin is as yet manufactured in this country."[70]

As the date for the meeting of the Board of Estimate approached, the *Herald* continued to push the theme of life against dollars. On December 18, two days before the Board of Estimate meeting, the executive committee of the *Herald* campaign met at the New York Academy of Medicine. The report by Biggs, Park, Prudden, Thatcher, and Brannan was read. While much lauded by the *Herald,* the report actually provided little new information. The history of the development of the serum and its proven efficacy and safety was described in cautious terms:

> In the hands of the eminent physicians and men of wide experience in this and other forms of disease in Europe, under the most careful systems of control which the representatives of the laboratories could devise, the new remedy has been authoritatively declared to be highly beneficent in a large number of cases of this fateful disease in children and also to be harmless.[71]

With none of the hyperbole that had characterized most of the *Herald*'s coverage, the report stated that the "determination of the exact protective and curative value" was yet to be established. The new remedy could save a considerable number of the children who died under current therapies. An emergency situation existed, they argued. Since the need for the new remedy was urgent and supplies limited, this alone was adequate justification for their work. The report ended with the promise that "if sufficient funds are provided, the emergency will be met in a safe and efficient manner." The signers' concluding motto was simply No Doubt of Success. Again the authors of the report suggested that there were two separate projects to produce the antitoxin, the *Herald*'s and the health department's. This was not the case.

From the outset, the timing of the *Herald*'s campaign and its relationship to the health department were suspect. Howson contends that the *Herald* campaign was orchestrated by the health department in order to influence the outcome of the Board of Estimate meeting.[72] There is some evidence to support this view. The content of the articles in the *Herald* emphasized the scarcity of the antitoxin. The impression was given that without the *Herald*'s fund, there would not be enough antitoxin to meet the needs in the city. Yet on December 14, Commissioner Edson reported that the health department would have sufficient antitoxin for the city from the fifteen horses already under treatment.[73] Edson stated in this same article that the *Herald* fund was an emergency fund and that the money from it should be made available to the city to purchase more horses. Though there is some discrepancy in Edson's figures, it appears from subsequent reports that the Board of Health would indeed have enough antitoxin to meet the needs of the city with the horses and equipment already in use.[74] The *Herald* consistently emphasized that its project was separate from that of the health department. In fact, there was only one project—the private effort to produce antitoxin initiated by Biggs, Prudden, and Park—which needed public funding in order to continue.

The only new element in the advisory committee's report was the pronouncement of an emergency situation in the city. Though both a scientific and humanitarian value were ascribed to their project, Biggs and Prudden clearly wanted to keep the focus on the issue of funding. They reported that while the number of cases of diphtheria in the city during December was higher than in previous years, "there is no epidemic in the city." It was the existence, but the unavailability, of the new remedy that now prompted the characterization of the situation as an "emergency." Such a declaration of an emergency was just the kind of language the *Herald* writers relished. In the days following the report, the *Herald* chastised New Yorkers for withholding their support: "It is practically criminal negligence to withhold support to the movement which means so much to the poor."

Money for Antitoxin

On Thursday, December 20, the Board of Estimate held hearings on the final appropriations of city funds for the next year. Biggs presented the health department's case. At the outset of the meeting, he produced a letter from Governor Flower recommending that the board grant a total of $52,000 for a bacteriological laboratory to "prosecute the work of preparing antitoxine."[75]

Biggs's argument was essentially the same one he had made in his letter to the board in August; he wanted funds for the purpose of the production of antitoxin and the scientific study of infectious diseases. First, he noted that the New York City Board of Health was unsurpassed in the world and that "Germany itself is adopting its methods in bacteriological research." Second, he again produced the statistics from the Municipal Council in Paris showing the decline in diphtheria deaths in that city: "The mortality has been reduced from 55 percent to 12 percent." Finally, he noted the scarcity of the new remedy: "There is only one way to get it, and that is to produce it, for which we must have an appropriation."[76]

Biggs won the day. An appropriation of $30,000 was granted. It was not, however, without stipulations. According to the *Times*, the mayor and the city controller said they "believed that the appropriation of the amount for actual experiments was a wise thing." Extra funds for the construction of a stable were denied. The $30,000 was "included under the head of the bacteriological bureau, but it was stipulated that it was to be applied for the special purpose of the preparation and use of antitoxine."[77] Biggs's attempt to get funding for bacteriological work other than the production of antitoxine was denied by the board. He had hoped to win support for the scientific study of infectious diseases more generally by demonstrating that diphtheria antitoxin was the product of "scientific research." Yet the city's leaders were not prepared to offer financial support for such an idea. As politicians they responded to public opinion. The focus of both the *Herald*'s and the *Times*'s coverage during the fall was on the antitoxin, which, though a product of scientific research, was viewed largely separate from its origins. The trade-off was that by using the newspapers to reach the public and translating the production of antitoxin into a "cause," Biggs at least got the funds he needed. Never one to be deferred by mere politicians, he moved on to his next plan, using again the new visibility of antitoxin to get increased funding for the bacteriological laboratory.

Conclusion

The *Herald* subscription campaign continued until November 1895 when nearly $7,500 was presented to the health department. Almost half of the total had been raised before the end of December 1894, during the height of the campaign. In contrast, both public and private contributions were substantially higher in France. The *Times* and the *Herald* continued regular coverage of the increasing use of antitoxin in the city during 1895. The *Times* reported most often on bulletins from the health department and physicians using the remedy. The *Herald* attempted to keep attention focused on its campaign but scaled down the size and extent of its coverage.

Newspaper reports constantly emphasized that antitoxin could not fail and was harmless; thus expectations were very high that all those on whom it was used would survive. The health department emerged at the end of the campaign with a more visible and powerful role. It controlled the production of the new drug and extended its authority over diphtheria, since mothers were urged to notify the Board of Health immediately at the first appearance of a sore throat in their children. Only these experts, they were told, could determine whether or not antitoxin should be used.

For physicians in the city, the newspaper focus on the new antitoxin had created a complicated situation. Few of them had access to the remedy, and therefore they had no firsthand knowledge of its efficacy. Still, no physician in New York City was unfamiliar with diphtheria, and most appreciated how complicated the disease could be to identify and treat. Many were still coming to terms with the bacteriological concept of the disease and with the health department's diagnostic program, which had only been in place for a year. Although many physicians took advantage of the diagnostic service, questions about it were slowly coming to the fore. The issue of interference by health department doctors in making diagnoses was already causing conflict by the fall of 1894. Nevertheless, with the most eminent and elite physicians in the city supporting antitoxin, physicians could not express doubts about it without appearing to be uninformed, unenlightened, or obstructionist. Whatever the physician's views, the newspaper attention had set expectations of the new remedy so high that it was virtually impossible for a physician to avoid using it if it was available. Although physicians were pushed to use antitoxin, their acceptance of it as the orthodox therapy in diphtheria would have to wait for a more formal medical evaluation.

Physicians' commentary had been used in the newspaper reports to validate the new remedy. No forum had been provided for them to express in detail their views about the implications of the new remedy for the treatment of diphtheria. This suggests that physicians' support of antitoxin at this time was more an expression of the desire to have the new remedy for evaluation than an unqualified approval of it.

The newspaper attention turned the production of diphtheria antitoxin into a public cause that the politicians controlling the city's funds had to address. Though Biggs's argument from August to December requesting funds for the production of antitoxin did not substantially change, the climate in the city toward the project changed dramatically in those months. As Mayor-elect William Strong prepared to take over City Hall in January 1895, the impetus for change in the usual way of conducting city business as practiced

under Tammany Hall was widely articulated. Reform groups in the city focused their attention on corruption in the police and street-cleaning departments. Investigations of these departments had led them to emphasize the
need for nonpartisan appointments in all municipal agencies. With respect
to the health department, it was largely the inspection of housing, food, and
the city water supply that came under scrutiny.[78]

In August 1894 there was no context for providing public money for the
production of diphtheria antitoxin in the city, regardless of the scientific
value of the new remedy. By December of that same year, the newspaper
reports had transformed the funding of diphtheria antitoxin from one based
on its little understood scientific value into an issue of public welfare.

For Biggs, Prudden, and Park, the newspaper attention given to antitoxin
was extraordinarily useful. Biggs, in particular, made himself very accessible
to the press. He was interviewed regularly and made his opinions known very
clearly and forcefully. Biggs was a very strategic and politically astute man.
He realized that his positions regarding public health were ahead of the majority of physicians and politicians in New York City. Yet he could not put
his ideas into practice through the health department without the support
of these groups and the public. He never failed to lobby behind the scenes,
often with Prudden, for the support he needed for his projects. The *Herald*'s
willingness to mount a public campaign to raise money for the antitoxin was
useful for him as long as he could be sure that it was his project that would
be supported. The composition of the *Herald*'s advisory board, obviously no
accident, virtually assured that his division would retain control over the production of antitoxin in the city.

The process of convincing the Board of Health and the Board of Estimate
of the value of the scientific work each time he had a new project was obviously not very efficient. Though he had successfully obtained funding for the
production of antitoxin, he had not gotten the support he wanted for other
projects despite the last-minute appeal from Governor Flower. Therefore,
early in the following year, Biggs submitted a proposal to the board to prepare
a bill for presentation to the state legislature that would amend New York
City's charter to allow the Board of Health to produce antitoxin for the poor
of the city *and* to sell any surplus produced. The receipts from these sales
were to be set aside in a special fund, known as the "Diphtheria Anti-toxine
Fund," for the sole use of the bacteriological division of the New York City
Board of Health.[79] The bill encountered no opposition in the city or the state
legislature and was passed in late spring 1895. Biggs was again able to capital-

ize on the favorable publicity associated with the antitoxin to obtain a fairly independent source of funding for bacteriological research in the city.

The *Herald*, meanwhile, had a campaign to capture the public's attention and promote itself as the champion of the poor. In the *Times* the antitoxin project found support through the paper's interest in supporting the best men, those with special expertise. The *Times* constantly derided the incompetence and patronage appointments made by Tammany Hall. Thus both newspapers promoted antitoxin because it served their interests in reaching the audiences they wanted to reach—and sold more papers as well.

4: THE DEATH OF
BERTHA VALENTINE

Given the climate generated by the *Herald*'s campaign by the end of 1894, physicians were hard pressed to explain why they would not use antitoxin to save a child from diphtheria. The health department claimed that use of the antitoxin led to dramatic drops in mortality from diphtheria, but physicians had little opportunity to make their own evaluation of the drug until it was more widely available in the early months of 1895. Physicians were thus put in the position of having to evaluate a drug that had already been declared the "right" treatment for the disease.

The first death following an injection of antitoxin in April 1895 gave New York City physicians an opportunity to discuss the drug.[1] As we shall see, there was a great deal of skepticism about the claims made for antitoxin. Advocates of the serum cast these skeptics as unreasonable and biased against bacteriology. Historians have tended to recount the position of the advocates rather than to examine the claims of the skeptics. This historical bias leaves the claims of the advocates unexamined while downplaying the very real problems associated with the introduction of antitoxin. Even when a medical innovation is generally accepted, it is important, as the historian John Pickstone has noted, to examine how its meaning and importance was contested and resolved among different professional groups.[2]

The debates over the therapeutic value of antitoxin reflected physicians' views of what constituted a proper source of knowledge.[3] Therapeutics, as John Harley Warner has argued, is both a cognitive system and a set of social practices that were central to the professional image and authority of physicians in the nineteenth century. Therapeutics are also a useful indicator of the changing real and perceived role of scientific knowledge. Thus the evaluation of diphtheria antitoxin was both a discussion of the efficacy of a new drug and a debate about the status of bacteriological knowledge versus clinical knowledge and the associated implications for medical authority and practice. The evaluation of diphtheria antitoxin commissioned by the leadership of the American Pediatric Society, as discussed in this chapter, illuminates the centrality of these issues.

The 1890s was a time of transition for medical therapeutics. Both physiology and bacteriology had brought new challenges to medical practice, but the depth of those challenges were yet to be revealed. This is especially true for the implications of bacteriology on clinical practice. The speculative period of bacteriology, when wild claims were made about new drugs, was coming to a close with the introduction of diphtheria antitoxin. It has been argued that physicians' acceptance of antitoxin reflected the acceptance of the bacteriological conception of diphtheria; therefore, historians have seen the acceptance of the new therapy as evidence of the triumph of bacteriology. But as we will see, physicians accepted antitoxin without being convinced that the bacteriological conception of diphtheria sufficiently encompassed the clinical realities of the disease. In addition, the debates over antitoxin explicitly addressed the professional authority of physicians, because antitoxin was introduced under the auspices of the health department, which immediately mandated its use among the poor.

Antitoxin in New York City

The first antitoxin produced in the laboratory of the New York City Board of Health was available for use in treating cases of diphtheria in January 1895. Immediately, two cases were treated successfully at the Willard Parker Hospital. Physicians were encouraged to refer cases to the health department's two inspectors, who would administer antitoxin upon request. By the middle of January, the *Medical Record* reported that the health department had approximately one hundred and fifty doses of antitoxin on hand and was manufacturing it at the rate of one hundred doses a week.[4] Though fewer than forty cases had been treated in New York City, interest in the remedy remained high. There had been a growing number of favorable reports about antitoxin from Europe as it became more widely available there.

The Death of Bertha Valentine

On April 1, 1895, an article appeared in the *New York Times* with the headline "Died After Taking Antitoxine." The article reported that seventeen-year-old Miss Bertha Valentine, of Brooklyn, had died ten minutes after an injection of antitoxin had been given to her by Dr. James L. Kortright, also of Brooklyn. Dr. Kortright asserted that the girl's death was undoubtedly due to antitoxin which he had purchased in the city. The girl had diphtheria, Dr. Kortright said, and five minutes after he gave the injection of antitoxin she had convulsions; in five more minutes she was dead. Dr. Kortright was unable to account for the strange action of the medicine unless some other preparation was put in the bottle by mistake. The article announced that an inquest would be held, though no blame was attached to the physician.

The death of Bertha Valentine was notable because it was the first recorded death attributed to diphtheria antitoxin in the city. According to an editorial in the *Times* that appeared on the day Miss Valentine's death was announced, "If it be true that the liquid used in this Brooklyn case was Behring's antitoxine, and that death was caused in less than ten minutes, it is the first instance of the kind, and the case is utterly at variance with a long and remarkable record of the experience of eminent physicians here and abroad."[5] Of course, the editorial commentary is somewhat overdrawn with respect to the American context. By April 1895 diphtheria antitoxin had only been available in sufficient amounts for a few months in New York City and the United States. The first use of it by the city's health department had only begun in January of that year, and private physicians who could obtain it had used it in a few cases during late December 1894.

The death of Bertha Valentine received extensive newspaper coverage. Both the *Times* and the *Herald* called for a thorough investigation of the case while assuring the public that "there is nothing to suggest that it [anti-toxin] is dangerous or that it can produce the effect which was seen in the Brooklyn case."[6] Physicians polled were divided on the cause of death. Some believed the antitoxin was contaminated even though tests of other samples from the same batch by the health department were found to be pure. Others thought that Dr. Kortright had failed to administer the serum properly. Of course, Dr. Kortright in his own defense rebuffed any notion that he had been negligent. In fact, he claimed to still be a supporter of the remedy in spite of his unfortunate experience, though he would use it with a bit more caution in the future.

> I shall continue to use it, but I shall observe the greatest caution. In order that I may not make another mistake, I shall first experiment in every case with the family cat. Every family has a cat, and I will inject a little of the serum into the cat before giving it to a patient . . . and if the cat continues in good health, I will feel safe in giving antitoxine to the patient.[7]

Though Kortright's comment is a bit glib, it does suggest that physicians had begun to use the antitoxin with somewhat limited understanding about how it worked.

On April 5, 1895, four days after the Valentine death, physicians met at the New York Academy of Medicine to discuss the antitoxin. It was noted that the large lecture room in which the meeting was held was inadequate to accommodate the largest and most representative audience that ever attended a meeting of the organization.[8] Several papers were presented that evening. Biggs gave the first paper, "Some Experiences in the Production and Use of Diphtheria Antitoxine," followed by Park's "The Preparation of the Diphtheria Antitoxin."

It was the Biggs paper that generated the most commentary. Biggs presented data on the use of diphtheria antitoxin in the city. First, he discussed data from cases treated by the sanitary inspectors of the health department in the tenement districts. In sum, 255 cases had been treated, with 40 deaths giving a mortality of 15.69 percent, compared to the previous year when the mortality among cases not treated with antitoxin was from 25 percent to 35 percent. He observed that there were higher mortality rates for those cases treated after the first day of the disease. Most of these cases ranged from moderate to serious in severity. In every case the diagnosis was established by bacteriological culture, so that, he asserted, "there can be no doubt as to the nature of the disease which was being treated."[9]

Biggs noted, however, less favorable results at the Willard Parker Hospital. At this institution, even with the antitoxin treatment the mortality rate was not appreciably different from that of the previous year. Cases there typically were seen in the later stages of the disease and often had serious complications. Biggs also felt that the less-than-spectacular results with antitoxin use in the hospital were due to the weakness of the serum used and that better results were expected with the use of higher dosages. Therefore, he emphasized two of the limitations of the antitoxin: that it had to be used early to be effective, and that it produced less favorable results in cases of diphtheria complicated by other infections. He also noted that there were side effects produced by the use of the serum, which included most commonly skin rashes and, in rare cases, erythema and swollen joints. He concluded with the comment,

> While there can be no question that occasionally cases of uncomplicated diphtheria will die even when antitoxin is administered early in the course of the disease, yet in my opinion, with the curative serum at our disposal, diphtheria, which has been the most fatal disease of infancy and early childhood, will be brought so completely under control, through the early administration of this remedy, that it will lose all of its terrors.[10]

Biggs's argument narrowly defined the boundaries of antitoxin's effectiveness while still ascribing to it the power to "completely control" the disease. He also downplayed the fact that the effectiveness of the new remedy was still being studied. He admitted that the first batch of antitoxin used at the Willard Parker Hospital was weak. Better results were obtained when stronger serum was used in February. The appropriate dosage for treatment was still unclear at this time.[11] Several contemporary authors have noted that the initial dosages used were too small to produce the most effective results in serious cases of diphtheria.[12]

After Biggs's presentation, Dr. Joseph E. Winters commented. Winters was the physician who led the attack against the antitoxin beginning with his comments at this 1895 meeting through the summer of 1896. Winters, who was clinical professor of diseases of children at City University and attending physician to the Willard Parker Hospital, had long experience with diphtheria in both private and hospital cases.[13] The designation of antitoxin by Biggs as the preferred treatment in the Willard Parker Hospital threatened to undermine his authority to dictate treatment there. His forceful attack against antitoxin underscores that both personal and professional interests were at stake in the debate over antitoxin.

Winters raised three important points in his comments to Biggs. The first was that there was no evidence that the formation of the pseudomembrane had been checked by the use of antitoxin. Second, recoveries from diphtheria at the Willard Parker Hospital using other methods were comparable to the results obtained with antitoxin. Third, he noted that there was a fairly mild variety of the disease prevailing in the city during the past winter. But most important, he argued that the statistics used to show the success of antitoxin were based on cases where there was no clinical evidence of disease. In his view, the last point was particularly damning. He noted,

> I should like to know why antitoxin is used in cases where the presence of the Klebs-Loeffler bacillus is the only evidence of diphtheria, when it is a well-established fact that the antitoxin treatment has no influence on the bacillus — it persists as long in cases treated with antitoxin as it does in cases treated without it.[14]

Winters felt strongly that cases of so-called diphtheria without any of the clinical manifestations of the disease should be eliminated from the statistical report, with the result that the death rate of cases under antitoxin treatment would be higher than what had occurred in the city during the previous year. He cited several cases that he had observed at the Willard Parker Hospital that directly contradicted the data presented by Biggs.

Finally, Winters spoke directly about what he considered to be the cause of death in the case of Bertha Valentine. He noted that proponents of the antitoxin had ignored the fact that individuals reacted differently to the action of remedies. "When this susceptibility exists, or to what extent was not something that physicians could know beforehand."[15] Yet it was the existence of this special susceptibility or idiosyncrasy that he believed had caused Valentine's unfortunate death. In conclusion, Winters came out forcefully in opposition to the antitoxin treatment of diphtheria because he had not found that it arrested the disease in cases with clinical symptoms and because the serum itself, he believed, posed an immediate danger to life.

Winters's comments spoke to the concerns physicians had about antitoxin. Did it arrest the symptoms of the disease they recognized as diphtheria? His comments and those of others speaking that same evening indicated that this point was still an issue. He explicitly rejected the view that diphtheria was defined solely by the presence of the bacillus. In his terms, a specific therapy for diphtheria had to have some effect on the symptoms of diphtheria. Furthermore, he was demanding that an assessment of antitoxin had to account for the changing incidence and virulence of diphtheria.

Biggs responded briefly to some of the charges made by Winters. First, he argued that it was manifestly untrue that a milder type of diphtheria had existed in the city during the winter. He also noted that he had not claimed that antitoxin could do anything other than neutralize the toxin of diphtheria when used early in the disease. Its impact on the pseudomembrane (at late stages in the disease) and on other constitutional symptoms was minimal. Second, he tried to defend his statistical data by noting that they were made up of cases reported to the health department where the diagnosis had been confirmed by bacteriological examination. He did not, however, concede that bacteriological data was in any way in conflict with clinical observations.

At the heart of this confrontation between Biggs and Winters were their differences over the conception of diphtheria. Winters was, in this case, critical of a kind of reductionist bacteriological conception of diphtheria that gave little consideration to the fact that diphtheria varied according to a range of variables. Cases might be mild or severe; they might involve the nose or throat; they often occurred with complications, especially in hospital settings.[16] Thus his criticisms were based on his experience of the complexity of diphtheria and his lack of faith in treatments that focused only on an unseen germ while ignoring the individual patient. The observation of the patient was central to his therapeutic decision making. His views were certainly understandable on the whole to many in the medical community, and his criticisms of the statistics from the Willard Parker Hospital were taken quite seriously by many. Contrary to the picture painted by Biggs's and Park's biographers, Winters's views were not dismissed out of hand by the medical community.[17] Their dispute was widely discussed in New York medical circles.

An editorial in the New York Medical Journal called the discussion at the Academy of Medicine a debate between two quite reasonable positions. On the one hand, there were those who believed that the remedy was useful and that they were "bound in conscience to use it." On the other hand, those who did not accept it were under just as much obligation to protest. The editor noted that until the unexplained death of Bertha Valentine and Winters's criticisms, it appeared that antitoxin was "destined to meet speedily with acceptance." Following the meeting at the Academy of Medicine, however, he believed that more experience with the remedy was needed before a final judgment was made.[18]

Park had his first opportunity to enter this discussion and respond to Dr. Winters's charges in June 1895, when he spoke before the annual meeting of the Massachusetts Medical Society. He opened his remarks by acknowledging to his audience the difficulties with diphtheria: "Diphtheria is one of the most difficult of diseases in which to judge the effect of treatment, for the different cases vary so enormously in the amount and in the location of their lesions." Though bacteriology had made some useful inroads in the study of diphtheria, he conceded that "most important factors, such as the virulence of the diphtheria bacilli, the extent to which other bacteria are associated with them, and the local and general susceptibility of the patient, are, as a rule, largely unknown to us." He continued by describing cases of diphtheria in hospital settings before the use of antitoxin. In his typology, mild cases were those in which the pseudomembrane was moderate in amount and confined to the area of the tonsils. These cases did well under any form of treatment unless there were complications. Dangerous cases were those where the pharynx and nose were also implicated or where the pseudomembrane extended below the larynx. In these cases death occurred either from toxic poisoning or from broncho-pneumonia or a septic bronchitis. There was also a class called septic, in which the diagnosis could only be assessed from the general condition of the patient rather than from any local lesion. One other feature of diphtheria cases was also important. They often began, as Jacobi had noted, with a small patch on the tonsils, followed by an increase in the pseudomembrane that no treatment was known to prevent.[19]

Thus having established the range of symptoms that characterized diphtheria cases, Park proceeded to address the issues raised by Winters, who had by this point published in pamphlet form the remarks he made at the April meeting of the New York Academy of Medicine. Park noted that Winters's widely quoted charges were deterring many around the country from using the antitoxin. Winters's comments were seen as credible because of his reputation and because he had the then rare experience of studying the effects of antitoxin in a hospital setting.

Park admitted that there were limitations to antitoxin treatment. It only worked against the poison produced by the diphtheria bacilli, yet as some critics claimed, in practice diphtheria often appeared with complications. It was difficult to determine how to classify these mixed infections. Park acknowledged that the indefinite conception of diphtheria still plagued the treatment of the disease and threatened to undermine the evaluation of antitoxin. "Here we see at once," he noted, "a limitation in the cure of the complex disease called diphtheria, for it is as correct, for instance, to class some of the cases met with as pneumonia complicated with diphtheria as to call them diphtheria complicated with pneumonia."[20]

If diphtheria with pneumonia could still be classed as diphtheria even by Park, then the problem of nomenclature verged on being insurmountable.

In addition, in some cases, after a certain degree of poisoning had taken place, the antitoxin had no effect; therefore, one could not expect cases seen late in the disease with serious complications to respond to antitoxin treatment. And finally, though in hospitals more children survived the disease after being given antitoxin than before, the results were less than satisfying, he noted, "but it seems as though still more should live."[21]

On the other hand, the use of diphtheria antitoxin in immunizing children had been quite remarkable. In two instances, at the Nursery and Child's Hospital at and the Mount Vernon Branch of the New York Infant Asylum, immunizing doses of antitoxin had been given to children who were exposed to diphtheria. These children had not caught the disease, and the outbreaks were brought under control. In closing, Park was very clear that diphtheria antitoxin had a distinct curative effect in diphtheria, yet it would not cure all cases even if given early in the disease. Nor did it destroy the diphtheria bacilli. Park had presented a measured analysis of the use of antitoxin in the treatment of diphtheria, but he had only completely refuted Winters's charge that the serum was a danger to patients.

Park was only marginally more successful than Biggs in convincing his audience of the value of antitoxin. Both men were hampered in this effort by the fact that the rhetoric used in soliciting the support for antitoxin during the previous winter had virtually ignored the complexities of the disease. When they were confronted with questions as physicians began to use it, they were forced to define its effectiveness in the narrowest terms to physicians who were still grappling with the unresolved problem of identifying what diphtheria was. Physicians were most concerned with the healing properties of antitoxin, and its limitations in this were not obviated by its more striking success in protecting children who were exposed to the disease. In fact, this last point, that antitoxin was useful in immunizing children against diphtheria, was barely acknowledged by physicians because of their overriding concern with treatment.

The discussion of the efficacy of antitoxin moved into the national arena when in May it was the main topic of discussion at the annual meeting of the prestigious Association of American Physicians. These physicians readily acknowledged that the public interest in antitoxin was in fact complicating their ability to determine the value of it in the treatment of serious cases of diphtheria. William Welch acknowledged that the press attention to the new remedy had led many more parents to take their children with mild sore throats to places where antitoxin was administered. Physicians were responding to this pressure, he asserted, and as a result more cases of diphtheria were being treated at an earlier stage in the disease than previously.[22]

Given parental pressure and the indefinite conception of diphtheria that persisted even as bacteriological diagnosis was more widely used, clinicians were put in a difficult position in trying to determine how to evaluate antitoxin. Mortality statistics from European hospitals were seen as being only suggestive of antitoxin's effectiveness, not definite proof. American physicians tended to place their trust only in facts that had been seen in America and interpreted through American eyes. For clinicians, municipal and hospital statistics showing dramatic decreases in diphtheria mortality rates with the use of antitoxin only exposed the obvious problem between the pre- and postbacteriological conception of diphtheria. These data were less than persuasive because in the prebacteriological period, municipal statistics included many severe cases of throat inflammation that were not due to diphtheria. The inclusion of these data resulted in a higher mortality rate attributed to the disease. After 1894, when bacteriological diagnosis was more widely used, many mild cases not previously defined as diphtheria were included. These cases increased the total number of cases but decreased the mortality rate since most of these cases recovered. Hospital statistics were considered to be unreliable because of the varying conditions under which patients were seen and because of the confounding presence of other infections as a result of the incomplete isolation of infectious disease patients in many hospitals.

Advocates of bacteriology like Park and Biggs placed greater confidence in the European hospital and the Willard Parker Hospital statistics because they were data that were sustained by laboratory experimentation rather than solely by clinical observation. Furthermore, such data were based on a newly objectified concept of disease that was specific rather than symptomatic and as such was seen as universally applicable.

Winters's charges, however, were taken quite seriously by Park and Biggs. Writing in 1931 about these early days when antitoxin was introduced, Park revealed that in response to Winters's charges, he had conducted an experiment at the Willard Parker Hospital in 1896 in which alternate patients were given antitoxin, and the remainder were not. The test lasted six weeks. According to Park, the cases that received antitoxin did well, whereas the rest did badly. He wrote, "The difference in the outcome of the cases was so great that we decided to discontinue the observations. We believed that although we had lost a few lives by it, we had gained a certainty as to the value of antitoxin which we would not otherwise have obtained, and this enabled us to persuade members of the medical profession much more rapidly than if we had not carried out the experiment."[23] Park and Biggs's faith in European statistics obviously did not relieve them of the need to verify the value of antitoxin themselves in order to convince their critics of its value.

Clinical medicine, on the other hand, emphasized observation and the discovery of facts at the bedside. Therapeutic practices rested upon careful observation of the idiosyncratic characteristics of patient, environment, and disease. The source of authority clinicians most esteemed was "their own experience."[24] The question remained, though, as to how this experience was to be categorized and evaluated.

The discussion at the Association of American Physicians, which included some of the leading figures in American medicine, revealed that Winters was not alone in his criticisms of either bacteriology or the antitoxin. Welch pointed out that the evaluation of the antitoxin was complicated by the fact that many clinicians continued to reject the bacteriological conception of diphtheria and mortality statistics based on it. Abraham Jacobi suggested that the evaluation of antitoxin needed to encompass the entire range of diphtheria cases:

> We do not want hundreds, but thousands of cases, in different years and different seasons, in order to get a final conclusion. There have been seasons in which other treatments have been just as successful. I know of 40 cases of diphtheria in which the bacillus was found, 39 of which got well one after another without antitoxin.[25]

Jacobi's suggestion, however, did not avoid the problem of identifying diphtheria. As Dr. A. C. Abbott of Johns Hopkins noted, "If we are to have statistics of any value on the subject, we must have some fixed point to go on, and the standpoint taken by the bacteriologist is the only one that supplies it."[26] Jacobi challenged this position:

> I want to place myself on record here as a heretic, we have been too much in the hands of our friends the bacteriologists. . . . We should be very poorly off if we should have to rely always on a culture to be made before making a diagnosis. I am not prepared as yet to accept the postulate, for it is nothing else but a postulate, that the diagnosis of diphtheria must necessarily include the presence of a bacillus.[27]

Jacobi was caught between his fear that bacteriologists were encroaching too closely on the physicians' authority with respect to diagnosis and his pragmatic therapeutic approach. He favored antitoxin treatment in "bacillus cases," as he called them, but as he said in this forum, "I cannot accept that we have to deal with diphtheria only when we find the bacillus."

The physicians at this meeting were divided over how the antitoxin could be evaluated. Some argued that statistics generated from long use of it in a variety of clinical cases could settle the matter, whereas, as Welch noted, for some, statistics in and of themselves would never be sufficient proof of antitoxin's efficacy. As to the matter of how to define diphtheria, most present at this meeting conceded that a very real split existed among physicians on this issue. The split between the bacteriologist's belief in laboratory facts and the clinician's need for experiential facts was evident at the close of this meeting. As Welch, supporting the bacteriologist's view, summed it up, there was ample experimental evidence in support of antitoxin, and the statistical evidence supported its efficacy. As far as he was concerned, antitoxin exerted a specific effect on diphtheria, and it was the duty of the physicians to use it.[28]

Despite the supportive position on antitoxin taken by Welch, others in the elite group in the Association of American Physicians and in the larger medical community as well continued to debate the efficacy of antitoxin through most of 1895 and 1896. Skepticism was still high as Dr. Floyd Crandall, editor of the *Archives of Pediatrics* expressed in the summer of 1895:

> *It seems to be the generally accepted opinion of observers, that antitoxin has a more or less decided effect in controlling the progress of diphtheria. It also seems clear that it produces in some cases unfavorable symptoms. It cannot, therefore, with our present knowledge, be unreservedly accepted or unreservedly condemned. The matter has resolved itself into a question whether these disadvantages are sufficient to counteract the advantages, and condemn the treatment.*[29]

The Evaluation of Antitoxin

By the summer of 1896 many physicians in New York and around the country had almost a year's experience using antitoxin. Published reports of its use in private cases appeared more and more frequently in the medical press. Statistics from hospitals around the world were published as well. For many, however, this seemingly endless discussion of cases and the dizzying lists and tables of statistics failed to provide a clear picture about how well antitoxin worked in the treatment of individual cases of diphtheria.

New York City physician Dr. W. H. Thomson amplified Jacobi's suggestion on how clinical experience might be used as a basis for evaluating antitoxin. He noted that there was considerable uncertainty in evaluating the treatment of any infectious disease. This uncertainty arose from a number of factors. First, each infectious disease ran a specifically defined course "naturally ending in recovery without treatment of a greater or less majority of those attacked by them." Second, there was the problem of diagnosis. Third, the entrance of germs into the body resulted in varying states of disease from mild to severe. In addition, pathogenic microbes themselves changed in their degree of virulence. Finally, with respect to diphtheria, mixed or polymicrobic infection was common. Given these factors he argued that "decisions on therapeutic questions, including the use of antitoxin, must rest on the question of comparative success, and on that alone."[30]

This was an important point, since many opponents of antitoxin had argued as if cases of the remedy's failure were sufficient to establish that it was of no value. Thomson made the point that "failures of a remedy are per se, no proof that it is not a remedy." Therefore, given that diphtheria was a widespread disease, the verdict on any remedy for it could only be ascertained by comparing the results of numerous competent observers from different localities and nationalities so that all personal and local influences could be ruled out: "In questions of therapeutics, personal opinions or supposed personal experience are so beset with fallacies that errors can be minimized only by the calculus of very large averages, and if the elements for these are not yet enough we must be content to wait until they are."[31] Thomson's proposal was that a comparative analysis of a large number of cases was the only way to overcome the problems of making an evaluation on the basis of the clinicians' personal experience. The hope was that the collective experience of clinicians could stand as evidence against the experimental evidence of bacteriology.

Collective Investigation

In May 1896 the American Pediatric Society commissioned a committee to investigate its use in private cases. Chaired by New York physician W. P. Northrup, the members of the committee included S. S. Adams, L. Emmett Holt, and Joseph O'Dwyer.[32] The committee decided to gather data from physicians in private practice rather than use hospital statistics largely because there were few cities in the United States with hospitals that accepted diphtheria patients. Furthermore, the conditions in those hospitals varied widely. Most importantly, the committee felt that hospital conditions differed so substantially from those found in private practice that they should be analyzed separately. The report examined 3,384 cases from 613 different physicians, demonstrating quite clearly that by the spring of 1896 antitoxin was being widely used in the 114 cities and towns in fifteen different states, the District of Columbia, and Canada polled by the Society.[33]

The report did not directly address the question of the identification of diphtheria, but the committee was obviously aware of the kind of data that would arouse controversy and thus were quite strategic in the kinds of cases they accepted for review. Two-thirds of the cases were confirmed by bacteriological tests, while the other one-third gave "pretty clear" clinical evidence of diphtheria. These were cases "that either had been contracted from other undoubted cases, or where the membrane had invaded other parts besides the tonsils, such as the palate, pharynx, nose, or larynx." Two hundred and forty-four cases in which the disease was confined to the tonsils and the diagnosis was not confirmed by culture "and therefore [was] open to question" were excluded.[34] The cases ranged from mild to the most severe. By discarding tonsillar cases that were not confirmed by culture, the committee eliminated the mild cases of diphtheria that were often at the center of disputes about antitoxin's effectiveness. And by including laryngeal cases, those which often required surgery, they could demonstrate the results from quite severe cases.

The cases evaluated in the report also included data provided by Biggs of 942 cases treated by health department inspectors in the tenement districts of New York and 1,468 cases from tenements in Chicago.[35] These data were obviously included as response to the ongoing debate in New York City. The overwhelming majority of the cases Biggs submitted were confirmed by culture. More important, they were described as serious rather than mild cases, and an unusually large number were given antitoxin several days after the onset of symptoms. In sum, the cases reviewed reflected both those in the early stages where bacteriological diagnosis was most useful, and the most severe cases where clinical diagnosis was most certain.

While the report did not explicitly comment on the problem of defining diphtheria, the inconsistency in physicians' use of antitoxin revealed that it was not based on the bacteriological conception of the disease. Many reported that they only used it when the condition of the patient had become alarmingly worse under ordinary methods of treatment. In addition, the authors commented that "many physicians being as yet in some dread of the unfavorable effects of the serum, have hesitated to use it in mild cases and have given it only in those which from the onset gave evidence of being of a severe type."[36] Several physicians reported that because of the expense of the serum, they tended not to use it in mild cases. The fact that many of these physicians also reported positive results is striking, given that they did not use antitoxin until the disease had progressed to the point of being serious. This hesitancy also suggests that they may have been reluctant to give adequate doses of it to their patients as well. In addition, despite the urging of people like Park that antitoxin be given in mild cases of diphtheria and given early in any case, physicians were clearly reluctant to follow this practice.

The three documented cases of death immediately following an injection of antitoxin, including Bertha Valentine's, were also examined, and none were attributed directly to antitoxin. The report noted that the type of diphtheria prevailing from 1895 to 1896 had not differed materially from that seen in previous years. This point assured physicians that the cases did not come from a season when diphtheria had been either unusually mild or virulent. While the report did not quite meet the standard that Jacobi had suggested of looking at thousands of cases under a variety of conditions and seasons, the fact that 1895 to 1896 had been a year when diphtheria was seen under its more normal endemic form obviated the need for testing antitoxin's effectiveness over a range of years.

The report came out favorably in support of the use of antitoxin in the treatment of diphtheria. The general mortality in the total of 5,794 cases was 12.3 percent. The most striking improvement was seen in cases injected in the first three days; among these the mortality was 7.3 percent. In the most severe cases of laryngeal diphtheria, the results with antitoxin appeared to far outweigh any other known method of treatment, reducing the mortality in these cases by half. The side effects of the treatment were noted and most were found not to be severe or consistent across the cases reviewed.

The summary of clinical comments published in the report were selected by the authors from those submitted, and it was noted that they were representative of the sentiments of physicians who sent in reports. Not surprisingly, they were uniformly in support of the serum. The remarks of Dr. Douglas Stewart of New York were quite striking:

My experiences in the past have been so unfortunate that the advocates of antiseptics or therapeusis were a constant surprise to me. It has been my fate to have the most desperate cases unloaded upon my shoulders. I have been forced into the belief that the profession was absolutely powerless in the presence of true diphtheria; have case after case of tube in the larynx and calomel fumigations at work. Previous to antitoxin, my only hope had become centered in nature and stimulants. In two years I have not lost a single case, and surely I may be pardoned if I suffer from diphtheriaphobia in a sub-acute form, and use antitoxin sometimes unnecessarily.[37]

Dr. Stewart's comments indicate how desperate physicians were to have something that worked in treating diphtheria, and this desperation certainly shaped their desire to try antitoxin. Dr. Thomson was right in his assessment that physicians would evaluate antitoxin by comparing it with the results they had obtained using other methods.

A final comment in the report was from a Dr. Reynolds of Baltimore. It was preceded by a cautionary comment from the authors that identified it as a case demonstrating the "dangers of relying too implicitly upon the bacteriological diagnosis." Dr. Reynolds had a case of a three-year-old male who died after antitoxin was administered on the fifth day. Reynolds claimed that the delay was due to the fact that the diagnostic report had only shown staphylococcus and streptococcus. A sister to the child subsequently contracted the disease, received antitoxin on the third day, and recovered. The reporter noted that he "would not wholly rely upon the culture test for diagnosis."[38] This final comment foreshadowed the committee's final views on the use of antitoxin and bacteriological diagnosis. It was recommended that antitoxin be used in all cases of diphtheria as early as possible on the basis of a clinical diagnosis, not waiting for bacteriological culture. As throughout the report, antitoxin was validated by the collective experiences of clinicians; whereas bacteriological diagnosis was put in the role of confirming clinical observation, not in supplanting it.[39]

The antitoxin report was unanimously accepted by the members of the American Pediatric Society:

It is the belief, however, of large numbers of experienced men that on the whole the benefits far outweigh the disadvantages, as to render the treatment not only proper, but morally obligatory upon the medical attendant. . . . It is certain that antitoxin has passed the stage of experiment; that it has secured a firm foot-hold as a therapeutic agent; that it is a powerful remedy in controlling diphtheria, and is here to stay.[40]

Conclusion

The question of how to evaluate the antitoxin had been resolved by the summer of 1896. The consensus of the medical profession was that the test should be made by evaluating the collective experiences of physicians in the context of private practice. By the end of 1896, diphtheria antitoxin had passed the tests put to it by physicians in the United States. It would be a mistake to assume, however, that no questions remained in the medical community. Historians in their evaluation of the success of the antidiphtheria program have too tightly coupled the two facets of the program, bacteriological diagnosis and the use of antitoxin. The acceptance of antitoxin as the standard therapy in diphtheria did not resolve the outstanding questions about the bacteriological conception of the disease or the value of bacteriological diagnosis. Physicians could accept the antitoxin, the product of bacteriological research, while not necessarily accepting the bacteriological definition of the disease. The point raised by Winters and others, that the bacteriological definition of diphtheria was problematic, still claimed the attention of physicians. As the editor of the *Medical Record* noted in commenting on the report, the tension between the bacteriological conception of diphtheria and the clinical view had not been resolved despite the success of the antitoxin treatment:

> There is no denying the fact that bacteriology pure and simple has appropriated more than its just share of credit in the present aspect of the question, nor is there any doubt of a desire among the very active workers in this now promising field of investigation to still more magnify the importance of their researches at the expense of clinical experience, upon which, after all is said and done, the practical use of every therapeutic measure must finally rest. Thus the tendency has very naturally shown itself to found the diagnosis absolutely on the bacteriological examination, with the result of greatly enlarging the number of cases and proportionately reducing the rate of mortality. . . . Bacteriology has made great advances in clearing up cases of reasonable doubt, but the broad generalizations regarding the true significance of a given microbe are not yet so firmly established as to be beyond the possibility of doubt or above the right of challenge.[41]

The tone of this editorial reflected the underlying concern evident in the collective report as well, that many physicians believed that advocates of bacteriology had overstated the value of their research to clinical practice. Physicians with long experience with diphtheria were willing to try remedies that they believed would do some good in the treatment of it. Some, like Abraham Jacobi, because he saw that antitoxin was useful could integrate it with his other treatments and gain even greater success. He remained deeply troubled, however, by the bacteriological conception of diphtheria. Writing in 1895 he said, "In regard to the nature of diphtheria. I believe, that after having observed it since 1858, I have only lived long enough to learn that I do not know what diphtheria is."[42] The fragmentation of diphtheria into a "true" and "false" form, and the unstable link between the presence of bacteria and the state of disease that bacteriological diagnosis had revealed, had not answered his questions about the disease despite his pragmatic incorporation of antitoxin.

The reaction to the bacteriological conception of diphtheria and the questions raised about antitoxin did not represent resistance to an amorphous "germ theory," nor was it merely a reaction to a threat of displacement from the bedside or a lack of concern with medical science as other historians have argued.[43] Physicians in New York City and throughout the country struggled to make sense of conflicting evidence about a new treatment applied to a disease that was one of the most difficult in which to judge the effect of treatment. They had to make this evaluation at the moment when the very basis of their therapeutic judgement, observation, and experience were increasingly being challenged by laboratory-based knowledge. They cautioned against a reductionist science that threatened to reshape their identity.[44]

Though the "germ" was no longer hypothetical as in the early days of bacteriology in the 1880s, attention to it still did not obviate the need to focus on the patient. Furthermore, the controversy over antitoxin was later viewed as being quite positive for the medical profession. It forced the gathering of more accurate statistics, brought more attention and study of antitoxin's side effects, and validated the profession's standards for therapeutics.[45] The hope of advocates of bacteriology in the New York City Health Department that diphtheria antitoxin would bring unqualified support for their program was not realized in this evaluation of the antitoxin. Antitoxin had limitations as well as striking benefits, but most importantly, in 1896 neither clinical nor bacteriological knowledge alone could explain all the facts about diphtheria. It still eluded being defined exclusively by the methods of either the clinic or the laboratory.

5: THE PROMISE UNFULFILLED

The bacteriological diagnostic program in New York City revealed more than just the problems involved in diagnosing diphtheria; it also made visible the sources of diphtheria contagion. The continuing prevalence of diphtheria in the city was traced to persons who harbored diphtheria bacilli in their throats and noses but had minimal or no clinical symptoms. There was evidence that these diphtheria "carriers" were sources of outbreaks of the disease in others. Public health officers used these data to reevaluate practices to control the transmission of diphtheria within families and communities. Regulations that demanded the isolation of those sick with diphtheria were extended to those who came in contact with the sick, and then extended further to those healthy persons who harbored the bacilli. Between roughly 1900 and 1920, the role of "carriers" of diphtheria increasingly occupied the attention of public health experts.

The New York City Health Department's diphtheria diagnostic program was the first municipal public health program to attempt to control the transmission of diphtheria through the identification and control of carriers. Park's 1895 study provided the crucial evidence that carriers were a source of diphtheria outbreaks, and data from the diagnostic program continued to support this view during the following years. Yet the extension of regulations to control cases of diphtheria to diphtheria carriers met with opposition from physicians and health officers and generated debates among observers around the country. These debates over the control of carriers exposed the limits of the bacteriological conception of diphtheria.

The control of diphtheria carriers revealed the problems with an approach that narrowed the focus of public health activity to the search for and elimination of germs.[1] Carriers could not be dealt with in the same way as germs found outside of the body. Thus the control of carriers also exposed unforeseen problems in the application of bacteriological knowledge in the public domain. Health officers found that neither physicians nor the public would accept restrictions on their activities solely on the basis of bacteriological knowledge. Public health practice reflected a more mediate approach than the laboratory evidence demanded.

Bacteriologists continued to maintain, however, that carriers were a source of danger and should be isolated from the healthy. They cited laboratory evidence showing that bacilli found in carriers were virulent and thus capable of causing disease in others. Yet this assertion was not as definitive as some bacteriologists claimed. There was also evidence that the properties of the diphtheria bacilli were unstable; thus it was not certain whether all diphtheria bacilli found in carriers were virulent.[2] Studies of the relationship between diphtheria and diphtheria-like bacilli raised a number of questions that bacteriologists could not answer. These questions had a direct bearing on the control of carriers. In the face of bacteriology's inability to precisely define the danger posed by carriers, health officers thus treated all carriers as dangerous. To confirm this danger by culturing a whole population, as bacteriologists suggested, called for a level of surveillance they felt was unwarranted by the evidence.

The debate over the control of carriers underscores two sets of issues. First, it shows one aspect of how the problem of bacterial variation was addressed by bacteriologists in a context in which the goal was the prevention of an infectious disease. The problem of the relationship between the varieties of diphtheria (and diphtheria-like bacilli) and the disease had been put aside in the wake of the successes of the diagnostic program and the use of antitoxin. The validity and reliability of the diagnostic program and the effectiveness of antitoxin were guaranteed by the assumption that the bacilli associated with diphtheria always exhibited the same characteristics. The presence of varieties of the diphtheria bacilli, virulent and nonvirulent, challenged the commitment to monomorphism that undergirded the diphtheria control program.[3] Second, public health practice demanded that bacteriological knowledge be generally applicable, reliable, and practical. Variation of diphtheria bacilli, and especially the question of the virulence of those bacilli, proved the limits of each of those criteria.

Bacteriologists and public health officers differed in their opinion of the "danger" posed by carriers. Bacteriologists equated the presence of virulent bacilli with danger. Public health officers characterized this view as extreme and turned to epidemiological evidence to define the danger that carriers posed in specific populations. For them danger was not merely the presence of virulent bacilli, but the potential of such bacilli to cause disease. By substituting epidemiological knowledge based on their observations of carriers for bacteriological knowledge, public health officers established the boundary of their professional authority and identity in the area of disease prevention.

As several recent studies attest, historians have not determined the full extent to which bacteriology limited public health practices to the finding and eliminating of germs, nor have they fully characterized the influence of public health practices on bacteriological research.[4] Little has been written by historians about diphtheria carriers because this issue has been overshadowed by a focus on the use of active immunization during the 1920s. Yet the problem of carriers and the absence of any treatment for them spurred important work on the study of mechanisms of immunity in the body. The failure to control diphtheria carriers served as an impetus both for laboratory researchers in developing methods to produce "artificial immunity," and for public health officers in considering the issue of "natural" immunity.

The carrier as both source and victim of infectious disease could easily have become the target of censure by the public or health authorities. However, the social perception of diphtheria carriers differed markedly from the infamous typhoid carriers. Unlike "Typhoid Mary," who became in this period the most visible symbol of the dangerous germ carrier, diphtheria carriers were not stigmatized.[5] In part, this was because diphtheria carriers were far more numerous than typhoid carriers; thus it was extremely difficult to determine whether one individual was directly responsible for an outbreak of disease. By their scarcity, individual typhoid carriers could be made the symbols of the anxieties and fears of disease germs in urban populations.

There were other practical differences as well. Typhoid carriers, once identified, could be rendered harmless to others by restricting their employment in food handling occupations, thus symbolizing the health officers' ability to control the disease. Such procedures could not be applied to diphtheria carriers because the infection was spread through respiration. In contrast, diphtheria carriers symbolized that there was simply no way to make the contagion visible, to keep it distant, or to render it harmless without imposing extreme methods of surveillance on an entire population. Diphtheria carriers were a persistent source of anxiety for health officers. They were the "spark which may or may not start the conflagration," and a constant reminder in this period that bacteriology had not, and perhaps could not, fulfill its potential to bring diphtheria under control.

The Carrier Problem

From the inception of the bacteriological program, the health department mandated that all persons who had diphtheria or who had been in contact with an active case be isolated until culture tests proved that their throats were free of bacilli. In addition, all those found to carry the bacilli without clinical symptoms were to be treated as if they were convalescent cases. In 1896, the health department responded to continuing complaints from physicians by changing its regulations regarding the isolation and quarantine of diphtheria cases. In private houses, the new regulations allowed the physician to determine how long isolation or quarantine should be maintained after all clinical symptoms had disappeared, but restrictions barring children from schools and churches until cultures showed the absence of diphtheria bacilli in the throat remained in place. When the physician released the patient from isolation, the health department was to be notified in writing so that disinfection of premises could be carried out. Teachers and others whose occupations brought them into immediate contact with children were not covered under the new regulations, nor were persons living in tenements, boarding houses, or hotels.

The new regulations were issued along with a circular outlining the health department's position on the transmission of diphtheria bacilli. It emphasized that healthy persons with diphtheria bacilli in their throats could convey the disease to others while they themselves remained well. In addition, the circular noted that during convalescence, diphtheria germs could persist in the throat for as long as three weeks in 30 percent of cases; four weeks or longer in 15 percent; and in about 5 percent of cases, five weeks or longer. Despite the paucity of research on this issue, it asserted without qualification that "experiments have shown that almost invariably these germs are virulent and capable of inducing the disease in others, so long as they persist in the throat, and persons having such germs in their throats may convey the disease to well persons at any time."[6] It was argued that these healthy persons carrying diphtheria bacilli in their throats posed a great risk to children. The circular strongly advised that these healthy persons remain completely isolated, though the health department could *enforce* isolation only on adults whose occupations brought them in regular contact with children.

Many physicians welcomed the new regulations because they felt that the former regulations were too arbitrary and sometimes encroached too closely upon the personal liberty of their patients. The stricter regulations had also given physicians a handy excuse to avoid using cultures. Medical commentators noted that "there was a growing tendency to postpone the sending of cultures through fear that patients who might suffer from very mild attacks of diphtheria might be made prisoners for days or weeks after every symptom of the disease had disappeared."[7]

Physicians obviously still did not share the health department's view that mild and convalescent cases of diphtheria posed a public danger. The editor of the *Archives of Pediatrics* expressed skepticism at the absolute certainty with which the health department asserted that the danger from such cases was real:

> It was questioned whether the Health Department is warranted in depriving citizens of their liberty unless they were able to say that the bacilli discovered in the throat, weeks after an attack of diphtheria, were virulent. It seemed doubtful to many men of large experience whether the scientific evidence yet available warranted such rigid measures.[8]

Again, physicians cast their own experience against the mandates of the health department. The editor reasoned, as would subsequent observers, that the most important issue was not simply the presence of diphtheria bacilli in convalescents, but the virulence of the bacteria they harbored. Physicians knew that it was very difficult to get rid of the bacilli in the throat after an attack of diphtheria. No effective method of accomplishing this had been found. Yet they were bound by the health department's regulations to keep such cases under observation for many weeks.

The health department recognized the opposition to their views and changed its regulations to allow the attending physician to assume any responsibility and risk that could arise once they released their patients from quarantine. The relaxation of the rules thus suggested that the health department was willing to accommodate the concerns of the physicians in order to ensure their participation in sending in cultures. Yet Biggs continued to assert that convalescent and mild cases of diphtheria were dangerous. He viewed physicians' resistance as another example of their refusal to accept the bacteriological conception of the disease.

The health department regulations were based on the work of Park and Beebe, who in their 1895 investigations had looked at the question of the transmission of diphtheria. Their report had noted a number of sources of infection: the pseudomembrane; discharges from diphtheria patients; the secretions of the nose and throat of convalescent cases in which the virulent bacilli persisted; and discharges from the throats of healthy individuals who had acquired the bacilli from being in contact with others having virulent germs on their persons or clothing. In the final case, they noted that "the bacilli may sometimes live and develop for days or weeks in the throat without causing any lesion."[9] In addition, they found that virulent diphtheria bacilli were apparent in about one percent of the healthy throats in New York City at the time of their examinations. While most of those cases had been in direct contact with cases of diphtheria, a few had not, as far as they could determine. Therefore, they recommended that members of a household in which a case of diphtheria existed "should be regarded as sources of danger, unless cultures from their throats show the absence of diphtheria bacilli."[10]

Park and Beebe had based their recommendations on the evidence gathered from cultures taken from the throats of 48 healthy children in families where they had been exposed to a case of diphtheria, and from 330 healthy persons in whom no contact with diphtheria could be determined. In the cases of the children exposed to diphtheria, they found bacilli in half of them. Forty percent of these later developed the disease. In the 330 persons with no exposure to diphtheria, they found virulent bacilli in eight persons, two of whom later developed the disease. Their evidence was drawn from close observations of outbreaks in families. For example, they cited the following case:

> In a family of 8 children 1 child sickened with diphtheria and a second child, a baby, was sent to a neighbor. The next day cultures showed this baby, as well as 2 other children, all of whom were apparently healthy, were infected with diphtheria bacilli. The 3 apparently healthy, but infected children, as well as the sick one were at once quarantined, but already 1 of the family to which the baby had been sent had contracted diphtheria from it.[11]

Other examples of outbreaks in isolated communities or tenements where the connections between each case were charted supported their view that seemingly healthy persons could harbor virulent bacilli after being exposed to an active case of the disease.[12] This recommendation implied that all cases of diphtheria were caused by another identifiable case or someone in the vicinity who carried virulent bacilli. Yet later in the report, Park and Beebe noted that when they examined their map of the city showing where diphtheria cases were located, they found that even with the most careful inquiry it was "impossible to find any connection with preceding cases of diphtheria in about one half of the first cases occurring in different houses." They had to admit that there were some elements in the transmission of diphtheria still unknown to them that "determine in every individual the occurrence or escape from diphtheria." Despite the widespread exposure to diphtheria in New York City, there were puzzling instances where they expected to find cases but did not. In sum, they found that it was simply impossible to trace the route of infection in many cases.[13] Despite this unresolved point, Biggs claimed that the health department's data were so numerous that there could be "no doubt now as the correctness of their observations," and furthermore, he argued, "it was perfectly evident that no means other than bacteriological examinations could determine the time of discharge of such patients from surveillance."[14]

In September 1896 Biggs directly addressed the question of virulence.[15] Citing Park's work and that of other European bacteriologists, Biggs argued that there was ample bacteriological research confirming that mild cases of sore throat were often the source of severe epidemic outbreaks of diphtheria. He emphasized that clinical support for such a view, noting Jacobi's views that mild cases of sore throat were often misdiagnosed cases of diphtheria, and that as such they were unrecognized sources of outbreaks of the disease. But since the medical community was still skeptical, Biggs ordered a set of experiments at the laboratory to attempt to "fix the exact value to be attached to the routine bacteriological examination in determining the true character of the cases described."[16] Taking a set of cultures from forty-eight cases in which physicians had given an uncertain diagnosis, Park and assistant bacteriologists Anna Williams and Alfred Beebe conducted virulence tests on guinea pigs.[17] A subsequent history of each case was also obtained in order to evaluate the results. Only three of the cultures showed nonvirulent bacilli. Though the sample size was small, Biggs concluded that this research, along with the earlier studies, proved conclusively that "when morphologically typical diphtheria bacilli are found in cultures made after the rather rude methods employed in diagnostic work, the result of the examinations may be generally accepted as indicating the existence of virulent diphtheria, unless animal tests have shown the organisms to be non-virulent."[18]

For Biggs the matter was essentially settled. As far as he was concerned, this study proved that he was correct in assuming that all bacilli found in mild cases of diphtheria were virulent and that therefore such individuals were capable of causing disease in others. More important, he assumed that all "true" diphtheria bacilli were virulent. He therefore divided diphtheria cases into four categories: first, those healthy persons who had been in contact with an active case of the disease (though he was not arguing that these persons had the disease); second, cases of simple sore throat with virulent bacilli present; third, cases presenting the ordinary clinical features of diphtheria plus the characteristic bacilli; and fourth, cases of sore throat with a false membrane but with no diphtheria bacilli, which were not diphtheria as defined bacteriologically, although such cases might be defined clinically as diphtheria. Biggs thought that this last category was still called diphtheria by some physicians because of their continuing resistance to the bacteriological conception of the disease. Any concession to such resistance was no longer acceptable to Biggs. He ended this report quite forcefully, asserting:

> It is this conception which I particularly wish to combat, and I desire especially to emphasize the statement that all inflammations of mucous membranes, due wholly or in part to the Klebs-Loeffler bacillus, should be included under the name diphtheria, without reference to the site or extent or intensity of the inflammatory process or the character of the exudate (emphasis mine).[19]

Despite Biggs's impatience with such views, physicians continued to raise questions about the bacteriological evidence. Dr. Samuel Meltzer, who later became a leading researcher at the Rockefeller Institute, took exception to Biggs's assertion that cases where the bacillus was not found were not to be considered diphtheria.[20] In his view, the New York City statistics were skewed because they relied only on a single culture to determine whether or not a case of diphtheria existed. Single cultures could be unreliable, he argued: "If the bacillus is not found in the first culture, the case is treated as a non-diphtheritic one, which seems to me a very hasty conclusion."[21] Countering Biggs, Meltzer argued that two points were required for establishing a case of diphtheria: a local lesion and the bacillus. Biggs, in response, reiterated that the pseudomembrane was not the definitive sign of diphtheria:

"There must be an inflammatory process produced by them [the bacilli] to constitute diphtheria . . . diphtheria is an inflammatory process of the mucous membrane produced by the diphtheria bacillus, whether there be a membrane present or not."[22] He insisted, though, that cases with no evidence of throat inflammation be treated as if they were diphtheria. Abraham Jacobi strenuously objected to this point on the grounds that the isolation Biggs demanded was unduly harsh:

> It has been a question with me whether the New York City Board of Health was correct in claiming that every case in which there was a Klebs-Loeffler bacillus in the mouth, even in otherwise healthy persons, should be isolated and prevented from going about and attending to business. I am of the opinion that is unnecessary. I believe that there are many here who have diphtheria bacilli in their mouths without ever having diphtheria or being in danger of developing it. A healthy mucous membrane is not infected with diphtheria though a bacillus is on its surface. I must say I have frequently looked upon this practice of the Health Board as unnecessarily cruel.[23]

Jacobi did not believe that healthy persons caused diphtheria or were in danger of developing diphtheria. He had consistently maintained that the normal, healthy mucous membrane could not be infected with the diphtheria bacillus. He, like Biggs, emphasized the necessity of the presence of some inflammatory process however mild.

Biggs did not concede Jacobi's point that the practices of the health department were cruel, but he did admit that such cases posed a serious problem for public health officials: "The question of the sanitary management of cases of diphtheria after convalescence, while the bacilli persist in the throat, what is to be done with such cases, this constitutes one of the most difficult problems in sanitary work and one which has not yet been solved."[24] Biggs was not alone in his view of the difficulties posed for public health officials in dealing with healthy diphtheria carriers. The problem called into question the very definition of diphtheria.

Dr. Hibbert W. Hill, director of the Massachusetts Bacteriological Laboratory, framed the problem in terms of the unstable boundary between the definition of diphtheria and "not diphtheria." In a paper entitled "The Official Definition of Diphtheria" presented at the meeting of the American Public Health Association in 1898, he wrote, "Perhaps no disease has been more discussed, certainly no disease is better known and understood, yet one hears again and again the question asked, where shall the line be drawn between diphtheria and not diphtheria?"[25] The problem as Hill presented it, was to decide whether diphtheria was patient + diphtheria bacillus or patient + diphtheria bacillus + typical lesions or patient + diphtheria bacillus + lesions due to the diphtheria bacillus. While the last definition was the most popular one, physicians and the public remained confused.

Hill explicated one source of the confusion—the divergent interests between physicians and public health officers who had to deal with the implications of the definition. For physicians, the recognition of the pathological changes caused by disease and the determination of their cause was the most important matter. To the health officer, on the other hand, the individual patient's diagnosis was always less important than the protection of public health. Hill correctly explained that the presence of the diphtheria bacillus was only an aid to diagnosis for the physician. If the patient did not have the disease, the physician's work was done. For the health officer, however, the presence of the bacillus was the critical issue. If the bacteria found was virulent, then the public had to be protected from this possible source of infection. "Where the bacillus is present, but the toxin absent or ineffective," he wrote, "it is the physician who is relieved of responsibility, whereas the medical officer's duty to the public remains to be performed."[26]

For health officers, then, the virulence of the bacteria was a crucial piece of information. Hill recognized the impracticality of performing virulence tests on every culture submitted to the health department. Such a procedure, even if feasible, would undermine the use of bacteriological cultures in diagnosis, because virulence tests took longer than twenty-four hours to perform. The health officer, then, was put into the position of assuming that all diphtheria bacilli detected were virulent in people who were in contact with a case of diphtheria since the experimental work, though not conclusive, tended to support such a position. Hill stressed that health officers had to insist that the condition "not diphtheria, but diphtheria bacilli present" was dangerous to the public health because of the opportunity afforded for the infection of others.

Hill correctly identified the carrier as a public health problem. Yet by requiring physicians to enforce isolation on those they considered to be well, Biggs's regulations pressed physicians to adopt the health officer's role. But as health officers well knew, the legal definition of diphtheria was based on the presence of a false membrane; therefore, if the definition was to be based solely on the presence of the bacillus, then existing laws would have to be modified. The public health definition of diphtheria might have been useful for the establishment of quarantine, but it was problematic for the maintenance of that quarantine. Dr. Charles V. Chapin, superintendent of health in Providence, Rhode Island, for example, had been subjected to heavy suits for damages when he had attempted to impose quarantine on persons with no clinical symptoms of diphtheria. Health officers thus had to determine the danger posed by both those exposed and those not exposed to diphtheria but who were incidentally found to have the diphtheria bacilli in their throats.

The New York City regulations required that quarantine and other regulations be based on the infective potential of the individual as identified by the presence of virulent bacilli, rather than solely on the presence of disease. Biggs's program was built on the view that the health officer could rightfully assume that the bacilli harbored in healthy throats was virulent. Park's laboratory, however, had devoted little attention to the question between 1896 and 1900. In her autobiographical account of this period, Anna Wessels Williams, Park's colleague in his laboratory work on diphtheria, reported that Park believed his 1895 work on this question had settled the matter:

> The question of diphtheria carriers had been brought to notice only a short time before. Dr. Park had been the first in this country to call attention to the possible danger from this condition, but he claimed that the virulent forms quickly disappeared from the throat after the case was cured. This practical viewpoint of the Director I found both satisfying and provoking—satisfying because of the calming effects on people who might be unduly frightened by an unnecessarily terrifying report, and provoking when I believed there should be more investigations on the subject. In the case of diphtheria carriers, Dr. Park thought it wasn't necessary.[27]

Williams was not alone in her concern that more investigation of this issue was needed. In July 1900, the prize-winning essay in the *Journal of the Massachusetts Association of Boards of Health* addressed the question, "What shall boards of health do officially with persons who are carrying diphtheria bacilli in their throats or noses without being ill, to prevent the spread of the disease?" The essay's author, Dr. Francis P. Denny, bacteriologist and lecturer in bacteriology at Harvard University, posed two questions: Were diphtheria bacilli found in healthy throats virulent? And how was this issue to be settled in practice? Echoing the concerns raised earlier by Hibbert Hill and others, Dr. Denny argued that such persons should be isolated until it was determined that the bacilli were virulent. If they were not, they should be released from quarantine.

The impracticalities of enforcing this position was at issue. Denny posed a hypothetical scenario: Suppose all cases of sore throat in which diphtheria bacilli were present were considered diphtheria and isolated, while healthy individuals with the bacilli in their throats were allowed to go free. How was anyone to decide what constituted a sore throat? There were obviously many individuals walking about whose tonsils were large and had reddened mucous membranes. Were only those who complained of sore throat to be isolated and those who did not let free? "I fear that, when people learn that this distinction was to be made, many of those in whose throats diphtheria bacilli were found would be unwilling to admit that there was anything the matter with their throats."[28]

In most cities by 1900, cultures were generally not taken from the throats of healthy individuals unless they were known to have been exposed to diphtheria. As Park and Beebe had shown and Denny had confirmed, the bacilli found in such cases were often virulent, yet these persons did not always develop the disease. The fact that this was still a point that few bacteriologists, much less physicians, understood was obvious from Denny's remarks.

Citing a German work that demonstrated that the blood of persons who had never had a throat infection contained a certain amount of antitoxin, Denny suggested that the answer might be that carriers had a natural immunity to diphtheria. Those convalescing from diphtheria were also found to have acquired immunity to the disease, though temporary. These individuals were thus protected from further development of diphtheria, even though virulent bacilli might continue to multiply in their throats for weeks after the disappearance of all symptoms.

Not surprisingly, Denny's explanation aroused little interest among the members of the Massachusetts Association of Boards of Health. This eclectic group consisted of several bacteriologists, physicians serving as health officers in the towns around Boston, representatives of large dairies, and physicians concerned with public health.[29] Many of these persons were unlikely to be familiar with the work Denny cited, but more importantly it appears that there was simply more interest in the practical aspects of the issue. In the discussion following the paper, Dr. Charles Chapin suggested that the crucial factor was the number of carriers in a community. He wanted to know how prevalent—that is, how widely distributed—diphtheria bacilli were in a given population. If there were only a few persons harboring diphtheria bacilli, then they could be isolated; but if there were hundreds or thousands, they certainly could not be. Furthermore, he argued, if the health officer only isolated the few he could find out about through the reports of physicians and not the rest, "then we are holding a very illogical position, and we shall soon be called to account, not only by the physicians and the public, but by the courts as well."[30]

Recalling the estimates made by Park and Beebe in New York City, Chapin argued that there could be as many as three to four thousand children in Boston that would have to be isolated, clearly an impossible task.[31] Dr. Brough, a Boston physician, emphatically decried the failure of such efforts even with convalescents:

> We have them constantly in the city, running over a period of weeks after the disappearance of the membrane; and practically, no isolation is carried out. They do not like to take the cases in the hospital, and we cannot release them; and it causes a great deal of inconvenience. It leads to a certain extent to a violation of the regulations of the board of health, which require people to isolate these contagious diseases; but it is almost impossible to make people understand that, when they have the bacilli in their throat, they are contagious. You may talk to them and tell them to isolate them, but they will not isolate them.[32]

This discussion illuminates the shifting lines of professional authority and expertise engendered by the public health attempts to control carriers. Those oriented toward bacteriology increasingly invoked laboratory evidence, while others focused on the practical implications of the evidence. Health officers sought a certain means or test to determine the extent of their authority in imposing quarantine and isolation on the so-called healthy. This led to the commissioning of a study on the issue.

A preliminary report was issued by the Massachusetts Association of Boards of Health in 1901, followed by a final report the next year. Then a subcommittee of the association, with Charles V. Chapin as chair, sent a questionnaire to various state and municipal boards of health in the United States and Canada in order to determine the percentage of healthy persons with diphtheria bacilli present in their throat or nose despite never having been exposed to the disease, and to assess the virulence of bacilli found in such cases. They also wanted to determine the degree of danger the cases represented, the best method for determining virulence, the number and site of cultures taken (single or multiple from nose or throat), the different methods of control used in these cases, and practices for controlling infected convalescents.[33]

These reports were the most comprehensive assessments of the problem of diphtheria carriers by health officers and bacteriologists at the time. The respondents included some of the leading public health officers and bacteriologists in the United States and Canada. The committee's conclusions and recommendations are worthy of note. Citing the investigations of Park, Chapin, and Denny, it concluded that in urban communities 1 to 2 percent of well persons among the general population were infected with diphtheria bacilli. And where healthy persons were exposed to diphtheria in families, schools, or orphanages, the number of those infected could range from 8 to 50 percent.[34] Only two of the seventy-four cities surveyed, Providence and Baltimore, attempted to systematically isolate healthy persons carrying diphtheria bacilli. The majority of responses, including a sampling of European cities, displayed decidedly mixed reactions on the practice.

The report concluded that only a small percentage of the typical diphtheria bacilli found in healthy persons not recently exposed to diphtheria were virulent. However, their respondents observed that the number of such infected individuals was so great that they represented "a very important" factor in the spread of the disease.[35]

Yet this ambiguous response, establishing that these individuals were an important source of infection, did not determine their danger for a population. The authors characterized this "danger" in terms that focused on the number of such individuals in a given population and on their individual behavior in specific environments; that is, their age, habits, surroundings, occupation, and intelligence. Children were particularly dangerous and endangered because of their habits and their susceptibility to the disease. Persons living in crowded tenements were dangerous because of their greater opportunity for contact with active cases. Milkmen and other food handlers, nurses, and teachers who worked with small children were also sources of danger. With respect to adult carriers, their knowledge of their potential to infect others and their willingness to take precautions were critical. By this schema, the population to be controlled was now narrowed from the general population to smaller groups identified by age and occupation.

The committee wrestled with the competing, conflicting interests of the public and infected individuals. It chose to take the most expedient course between the two: "The responsibility is largely shifted to the latter, and in the case of intelligent persons, the individual responsibility in disseminating disease should be clearly placed before the infected person."[36] At the same time, the committee encouraged health authorities to be vigilant and to be prepared at any time to isolate persons discovered with presumably virulent diphtheria bacilli.

These recommendations reflected the emerging tenets of the "New Public Health," which turned away from cleaning the environment and disinfecting sick rooms toward a focus on the human sources of infection. Personal hygiene was the watchword of the day. Individuals could protect themselves from communicable disease by fostering personal habits that limited their exposure to the secretions and excretions of people around them—both sick and well. The health officer's job centered on persuading carriers to accept responsibility for protecting others by voluntarily limiting their contact with them. Efforts to control diphtheria carriers were reduced to educating the carriers that they could detect.

The report did not acknowledge a basic flaw in such an approach: the only carriers who could be identified were those known to be in contact with an active case of diphtheria. All other carriers could not be identified without increasing the surveillance of entire populations. Many health officers working in increasingly diverse populations in urban areas had an appreciation of the negative implications involved with such surveillance.

Some officers remained troubled by the inadequacy of the solution to the carrier problem. Personal hygiene and increased surveillance were less than optimal methods of control. Surely, there must be a more certain way to determine the danger posed by carriers.

The bacteriological issues were addressed in the second section of the report. Dr. Theobald Smith, bacteriologist and director of the Antitoxin and Vaccine Laboratory of the Massachusetts State Board of Health and professor of comparative pathology at Harvard, authored the introduction to this section. Smith reiterated current knowledge that the persistence of diphtheria bacilli in the healthy was "entirely in harmony with our knowledge concerning the relation of disease germs to the individual liable to their attack." He understood that this was difficult for many to accept due to continued misconceptions about the germ theory of disease. The popular conception that disease germs entered, permeated, and then passed out of the body "like an orderly army without stragglers," in Smith's words, was not confirmed by observation or experiment.

Smith explained that there were two opposing forces in every infectious disease: the resistance of the individual and the virulence of the bacteria.

The resultant disease was thus a variable phenomenon. In a variety of diseases, persons who were to all appearances well could carry and transmit disease germs. In Smith's view, there was no basis for assuming that such bacteria were necessarily nonvirulent. That they were harmless to the carrier was simply an expression of the carrier's immunity, either inherited or acquired at some earlier period through exposure to the disease. Smith asserted that laboratory tests confirmed that "such microbes, when identified as belonging to one or more of the species of pathogenic bacteria, are as a rule, of the usual virulence."[37] He further explained that such organisms were frequently carried in such a way as to be easily shed, and hence to be a source of danger to others, as in typhoid and cholera. Given that the diphtheria bacillus tended to localize on the surface of the throat and nose membranes, sending its toxin inward, Smith argued that this localization could allow the bacilli a "special opportunity to escape destruction by the body and to vegetate in the body in immune individuals for a time." Thus it would not be surprising to find in the future even more persons carrying diphtheria bacilli than had previously been identified.

Smith's position was in direct opposition to many of the major conclusions of the report. Basing his argument on the process of infection by pathogenic bacteria in the body, he saw clearly that the danger posed by carriers of bacteria was significant and that as such there was no scientific basis to make distinctions based on whether or not the person had been recently exposed to the disease:

> As there is no sharp line to be drawn between the healthy and the diseased state, one shading imperceptibly into the other, so there is no sharp line to be drawn, for many infections at least, between the time when micro-organisms are still in the body and when they have all been destroyed or eliminated. It follows logically as shown by the work of this committee, that well persons may at times be the source of infectious diseases.[38]

As a bacteriologist, Smith could not support the recommendations made by the committee. Healthy persons carrying disease germs were sources of outbreaks of diphtheria:

> In the capacity of bacteriologists it becomes our duty to emphasize the facts, that well persons may be the source of virulent disease germs, and that public health authorities cannot concentrate their attention on health and disease as such, but must take cognizance first and foremost of the whereabouts of disease germs; and

in so doing they practically admit that the carrier of such germs, even if well, may be fitly compared to the spark which may or may not start the conflagration.[39]

The measures to be carried out in dealing with individual carriers once discovered, Smith conceded, must be left to the health officers to devise: "Such measures would naturally depend on the relative dangerousness of the disease involved, on the relative ease with which the bacteria is disseminated, and in fact, upon a number of conditions which do not come within the scope of the bacteriologist, but which must be settled on purely practical grounds."[40] While Smith's remarks indicate that bacteriologists, health officers, and physicians agreed that the carrier was a potential source of diphtheria outbreaks, these three groups were not in agreement as to how such persons should be treated.

Smith argued that all carriers of virulent disease germs could be "fitly compared to the spark which may or may not start a conflagration" and that bacteriological data gave ample support to this view. He did not believe that health officers should continue to try to separate out only those carrying virulent bacteria since all the scientific evidence indicated that such bacteria were usually virulent.

Public health experts took a different view. They did not dispute the facts presented by bacteriologists, but they sought some way to make distinctions among carriers in order to exercise control over these sources of infection. Bacteriological evidence in support of the practice of isolating bacilli carriers was of little help, since neither patients nor physicians would accept such evidence, even in cases of those known to have been exposed to a case of diphtheria. The authors wrote: "It is useless to try and tell him [the carrier] that the conditions are different in his case, that his bacilli are derived from a recent case, and have been proved virulent. He will not accept the reasoning, and very likely his medical adviser will not accept it."[41]

Physicians were not willing to accept that the healthy should be restrained based on bacteriological evidence. By 1902, public health officers already had some experience with trying to enforce isolation of healthy carriers of diphtheria and were being met with increasing opposition, especially when they attempted to restrain the movements of adults. In the end, as Smith noted, bacteriologists conceded that they could only present the facts and leave it to others to find a practical solution to the problem. Bacteriologists did not have the authority, and perhaps in this case the desire, to engage in a campaign to push their view of the danger carriers posed.

Diphtheria the Uncontrolled

Despite the Massachusetts Association of Boards of Health's assessment of the issue of diphtheria carriers, questions persisted among public health officials. By 1910, the term *carrier* was used more and more often to describe persons who harbored pathogenic microorganisms in their bodies. Though this phenomenon was recognized early in the bacteriological study of diphtheria, the term became more routinely used as researchers discovered the carrier state in other diseases. In 1893, Koch had identified carriers (*bazillenträgers*) as being the source of outbreaks of cholera. But it was his work on the role of carriers in the transmission of typhoid in 1902 that brought the greatest attention to the subject.[42] The term first appears as a category in the Index-Catalog of the Library of the Surgeon-General's Office in its third series published in 1923. By 1920, researchers found that carriers played a role in the transmission of a number of diseases including meningitis, pneumonia, and anterior poliomyelitis.

With the recognition of the fact that carriers played a significant role in the transmission of some infectious diseases came the need for greater specificity in the concept. German, British, and American researchers established different schemes for distinguishing between types of carriers, but in general two broad classes of carriers were defined. The first were "convalescent carriers," those who had recovered from an attack of the disease but continued to harbor the bacteria in the body. The second class were defined as "contact carriers," those who had never suffered from a recognized attack of disease but still harbored the bacteria after contact with an active case.[43]

In addition to defining the type of carrier, some distinctions also had to be made based on the duration of the carrier state. Thus carriers were referred to as either transitory, indicating that the carrier state was of short or intermittent length, or chronic, which suggested that a person consistently harbored bacteria over a significant amount of time (for weeks, months, or years). There were also persons called "healthy carriers" in whom no evidence of contact with an active case of disease could be found. In some cases, chronic and contact carriers were also referred to as healthy carriers in order to distinguish them from convalescent carriers.

This nomenclature was by no means fixed during the first decades of the twentieth century. The categories of carriers evolved and changed as both epidemiological and laboratory research uncovered more details about the phenomenon. The boundaries between the categories were fluid as well. For example, many debates centered on whether or not persons classed as contact carriers were actually convalescent carriers who had suffered from cases of disease so mild that they had gone unrecognized by both the patient and the physician.

The carrier thus came to symbolize what Theobald Smith had called the "line between the diseased and the healthy." Yet paradoxically, carriers were also both source and victim of infectious disease. In many diseases, including diphtheria, the carrier state could not be eliminated by any medical or surgical intervention, though the search for such interventions increased during the early decades of the twentieth century. Given the absence of interventions, the focus shifted to controlling the behavior of carriers.

As health officers continued to characterize all carriers in terms of their potential to produce disease, the differences between carriers were elided. Carriers were often described in the most pejorative terms as reflected in the words of one health officer who wrote: "While we accredit nature with marvelous adaptations for the welfare of mankind, it should not be forgotten that a typhoid gall bladder or a diphtheria tonsil represent a diabolical mechanism for the perpetuation of some of man's real enemies."[44] The use of such language made the carrier synonymous with the already stigmatized pathogenic bacteria that they harbored.

The public response to the concept of carriers varied widely, although according to physicians' and public health officers' accounts, the concept was not generally understood. An episode that occurred in Cambridge, Massachusetts, in 1901 gives some indication of the public attitude. In this case a man was taken ill with a sore throat, and a physician was called to attend him. Although the doctor did not consider his condition to be serious, he ordered a culture to be taken. The board of health report indicated that diphtheria bacilli were present in the patient's throat, so according to Massachusetts law, the house where the patient lived was placarded and the patient quarantined. The patient wrote to the editors of the *Boston Medical and Surgical Journal*, complaining that he had been unfairly treated:

> Of course I know, everyone knows, that culture taking is productive of most beneficial results. It assists the diagnostician and safeguards the public at large. But should it not be applied with some reasonable discrimination? At present, in Cambridge at least, a person's liberty is placed in the discretion of one man, who is privileged to act in the most arbitrary manner and from whose decision there is no appeal. Who, after

he has once played the monkey in a farce like this, will ever allow a physician to take
another culture from his throat, or if he does get one will ever let him live to lug it
away?[45]

Dismissing the threat of violence, the editors conceded that the position
taken by this man was understandable; they could sympathize with his feel-
ing that despite the scientific evidence, he was well. Physicians understood
that bacteriological knowledge, while it had expanded the notion of disease,
had not necessarily brought clarity to the notion of illness. Ordinary people
in such circumstances were not ready to relinquish their ability to determine
the state of their own health. The solution to the dilemma, the editors argued, lay in gathering more knowledge of the virulence of the diphtheria
bacilli under varying conditions, "a subject upon which we are quite aware
an enormous amount of work has already been done, but upon which still
more needs to be accomplished."[46]

Bacteriology and the Problem of Virulence

It was widely recognized among bacteriologists that many varieties of the
diphtheria bacillus existed. As Park had found in 1895, and as many researchers had confirmed, there were diphtheria bacilli as well as pseudo- or
diphtheria-like bacilli widely present in most urban populations.[47] Park had
given the name "pseudo-diphtheria bacilli" to a group of organisms with certain cultural and morphological characteristics distinct from those of the
diphtheria bacillus. Such bacilli were not found in typical cases of diphtheria, and no case of infection had been caused by them. For all these reasons,
Park considered these nonvirulent typical diphtheria bacilli to be true diphtheria bacilli that had lost their virulence.

Park's view reflected the prevailing view among bacteriologists that
disease-causing microorganisms had stable characteristics. This bacteriological doctrine, known as *monomorphism*, was a fundamental aspect of the
germ theory of disease.[48] The pathogenic qualities of the diphtheria bacilli
had been the focus of bacteriological research because of its relevance to the
disease. Only virulent bacilli produced toxin that caused diphtheria's symptoms. When the carrier problem emerged, the presence of other diphtheria-
like bacilli and their relationship to the disease became a research problem
for bacteriologists. Since Park's study, researchers had continued to investi-

gate whether there were several species of diphtheria bacilli or whether the diphtheria bacilli existed in different grades of virulence.

The varieties of diphtheria bacilli were a puzzle to bacteriologists, leading some to question the etiological significance of these bacilli to the disease. Researchers were divided on the issue. Some believed that diphtheria-like bacilli were simply variations of the typical diphtheria bacillus. These researchers held that both the diphtheria and the diphtheria-like bacilli had many unstable properties, their form, cultural characteristics, and pathogenicity all varying within a wide limit. Other researchers differentiated among these characteristics and argued that certain forms of the bacillus had such stable properties that they constituted a distinct species.

The question most often raised was whether nonvirulent forms of diphtheria bacilli could become virulent and vice versa. The answer to this question had obvious implications for the control of carriers. If virulence was not a stable characteristic, then the presumption that all carriers were dangerous was open to question.

Anna Wessels Williams reported on this question in 1902.[49] She had conducted a seven-year study of the different varieties of diphtheria-like bacilli in New York City in order to determine which varieties were true subspecies and the virulence of each form. Her samples came from cultures taken from clinically typical cases of diphtheria from a variety of sources: from patients at the Hospital for Contagious Diseases, from healthy and diseased throats in a town during an epidemic of diphtheria, and from the throats of children during an epidemic in a home for destitute children. She confirmed that typical diphtheria bacilli were a distinct species from the atypical diphtheria-like bacilli, the so-called pseudoforms. Most important, she found that the virulent diphtheria bacilli did not change into any nonvirulent diphtheria-like bacilli in the throats and noses of people during an attack of diphtheria or in healthy throats. From a public health standpoint, therefore, the nonvirulent forms of diphtheria bacilli could be regarded as harmless, since virulent forms were the only ones capable of producing infection.[50]

While Williams's work cleared up some aspects of the stability of typical diphtheria bacilli and confirmed the relative harmlessness of the diphtheria-like varieties, her work did not settle the question of virulence. Another researcher in Park's lab, Anna I. von Sholly, extended Williams's research by revisiting the work on the presence of diphtheria bacilli in apparently healthy throats.[51]

Von Sholly's paper raises some interesting points about why no consensus appeared on the question of virulence. In her review of the literature, some studies showed that the diphtheria bacillus was fairly common in the mouths of healthy persons and thus led many physicians to believe the condition was so common it could be ignored; other studies showed that such cases were fairly rare when there was no direct contact or exposure to diphtheria. Throughout the studies reviewed from 1894 to 1907, the extent to which typical virulent diphtheria bacilli were present in the healthy throats of persons never exposed to diphtheria was still unclear. The variable methods used in each study made comparison difficult.

> The results depend largely on what each individual investigator considers a "normal throat," how far he differentiates the members of the group of organisms showing the morphological characteristics of the diphtheria bacillus, on the environment of the cases examined, and the severity of the disease as it exists in the community.[52]

Von Sholly's review showed that a bewildering number of facts on this question by researchers had been accumulated from all over the world. Each of these studies had different definitions of the critical variables in the problem. Each of these variables — the "normal throat," the "typical diphtheria bacillus," and the "community" — had to be examined in order to understand the role of carriers in transmitting diphtheria.

Identifying the sources of infection in a city like New York was virtually impossible. There were 8,000 to 12,000 cases of diphtheria reported each year, and so many possible locations where one could be exposed: tenements, schools, shops, public conveyances, hotels, and restaurants. Thus, by necessity one had to assume that infection always came from a previously infected individual, whether it be "mediate or direct, remote or recent." Given the conditions in the city, von Sholly could not, as she wrote, conduct a study "to prove or deny the presence or absence of diphtheria bacilli in the healthy throats of persons never exposed to diphtheria" (emphasis von Sholly).[53] The best she could hope to do was to determine approximately where virulent diphtheria bacilli were found in the apparently normal throats of individuals among whom diphtheria was most frequent, namely children in crowded tenements and institutions.

Though she looked at a large number of diphtheria patients and contacts, there was nothing in von Sholly's study that contradicted the earlier work of Park, Beebe, or Williams. She found diphtheria-like organisms in persons where no direct exposure to diphtheria could be traced, and at least one-third of those persons harbored virulent strains. Virulent strains were found in those exposed to diphtheria four times more often than among the non-exposed.

Von Sholly's and Williams's work was typical of bacteriological studies on the virulence question in the first decades of the twentieth century. Park and his associates had shown that the "true" diphtheria bacillus was virulent and did not lose its virulence over the course of the disease. In both von Sholly's and Williams's studies, samples were taken from a wide range of sites throughout the city in a variety of institutions and were subjected to virulence tests in the laboratory. These studies largely confirmed Park's earlier conclusions. By the 1920s, however, he had begun to question some of his own earlier assumptions about the concept of virulence.

Initially, Park had defined virulence as the ability of bacteria to produce toxin, which he equated with the ability of carriers to infect others. At this later point, he argued that the ability of carriers to infect others was not necessarily directly proportional to the bacteria's toxin-producing capacity.[54] This key assumption had not been addressed in the studies conducted in Park's laboratory or by other bacteriological studies. Later, through his own observations of carriers and other epidemiological evidence, his views on the concept of virulence shifted.

Epidemiology and the Carrier Problem

Health officers had continued to raise doubts about the bacteriological research on carriers into the 1920s. As the major author of the Massachusetts Associations of Boards of Health report, Charles Chapin received a number of letters from public health officers around the country indicating that a high level of uncertainty about the evidence supporting the danger posed by diphtheria bacilli carriers still existed. Dr. William C. Woodward, writing to Chapin in 1901, noted:

> Your letter expresses my opinion relative to the status of diphtheria bacilli in determining the quarantine of patients suffering from that disease. My practice as Health Officer is, however, different. I have not yet felt justified in disregarding the opinions of practically all American health officers and imposing my own upon this community. I can see no reason whatsoever why diphtheria bacilli known to be virulent should be more dangerous to the community in the air passages of a person who has presented clinical symptoms but has gotten entirely rid of them than in the air passages of an individual who has never had such symptoms (emphasis Woodward's).[55]

Woodward believed that the danger from diphtheria bacilli in the throats of well persons had been "enormously exaggerated." His impression was that the number of cases of diphtheria had not been reduced in any place where carriers were routinely isolated. Woodward apparently was not convinced by the position taken by other public health officers. He agreed that it was not possible to quarantine convalescents if the healthy carriers were not quarantined as well. He was also correct in asserting that the extent to which carriers were responsible for outbreaks of diphtheria had not been determined by the Massachusetts Association of Boards of Health.

Chapin countered Woodward, arguing that the restriction of carriers based on throat cultures had reduced the prevalence of the disease, though he could not chart the extent of this reduction. It was almost impossible to evaluate whether the decline in diphtheria in some locales was due to better isolation, the use of antitoxin, or the natural wavelike variation in the intensity of the disease that had been charted for decades. Chapin saw the problem as the inability of health officers to employ a rigorous system of quarantine based on the use of cultures.

Regardless of his views, Chapin bemoaned the fact that after ten years of concentrated effort, diphtheria remained one of the most serious of the common infectious diseases:

> Notwithstanding that the diagnosis and the determination of the duration of infection are more certain than in any other disease, the application of the most stringent measures of isolation and disinfection, based on the most thorough bacteriologic examinations, do not appear to have been much, if any more effectual, than the very loose and uncertain methods practiced 10 or 15 years ago. We have been steadily strengthening our restrictive measures, but with little avail. We were justified in doing this so long as we were justified in the hope that it was possible to control all or nearly all the sources of infection.[56]

As Chapin noted, carriers were so numerous that it was simply impossible to identify or control them in an urban setting; however, public health officers continued to investigate the issue. There were numerous published reports on the number of diphtheria carriers found in various institutions, studies of how widely distributed they were in communities, discussions of the use of one or two negative cultures to determine the final disappearance of the bacilli, and studies to determine how many carriers harbored virulent bacilli.

Given that so many studies and reports had failed to settle the issue, the question was put to Chapin, "Are these really sources of infection, or are they, as so many believe, an imaginary danger evolved in the brain of the laboratory worker?" Chapin had no doubt that carriers were real and that the evidence overwhelmingly supported the view that they caused diphtheria:

> It is well again to emphasize the fact that, long before Klebs and Loeffler identified the bacillus of this disease, a few able clinical observers like Jacobi believed that evidence pointed clearly to the great danger of these unrecognized sources of infection. The bacteriologists have not raised the bugaboo of carrier infection, they have simply explained the facts which observing men have long recognized.[57]

For health officers, it was precisely their observations of carriers that caused them to question the bacteriological evidence. The carrier problem was so confusing that some even questioned whether the diphtheria bacillus was the etiological agent of diphtheria.[58] Despite the ever-increasing knowledge about the bacteriology of diphtheria and its mode of propagation, it was still a common infection. Why is this? they asked. Even though they recognized that they missed many carriers, they were surprised that they did not see more diphtheria in cities where the number of carriers was thought to be high. Writing about the situation in Washington, D.C., Woodward argued that if 2 percent of the community were carriers, then physicians should be seeing many more cases of diphtheria from contact with these persons than were reported.[59] The only explanation he could offer was the speculative claim that most carriers had such good personal habits that their danger to others was diminished.

Hibbert Hill questioned whether the infectiveness of the carrier was equal to that of those convalescing from diphtheria. Echoing Woodward, Hill also asked why no one had explained how carriers not having very intimate contact with any individual for any length of time could possibly be more dangerous than the convalescent who had close contact with household members. "If the infectiveness of carriers is equal to the convalescent," he argued, "then the amount of diphtheria in Boston ought to be higher." Nine years later, in 1910, he confessed that he still could not determine the role carriers played in diphtheria outbreaks. By this time he had relocated from Boston to Minnesota, and he saw the problem somewhat differently. In the western regions, he had to account for sudden outbreaks of diphtheria in cities where it was comparatively rare; whereas in the eastern urban areas, the problem was accounting for a few scattered cases and for institutional outbreaks. Carriers were responsible for the latter, he believed, but did not adequately explain the former.[60]

By 1925, epidemiological studies had generated a great deal of information about carriers. Park's views changed in 1923 when he observed twenty families for two to four-and-a-half months, in which one or more carriers of virulent diphtheria had been identified in each month. No case of diphtheria developed in any of these families over the period of study.[61]

A much more extensive study of five hundred cases in a section of Baltimore was conducted by epidemiologists at Johns Hopkins University. The study was designed to determine the risk of attack for carriers' families and to compare this risk with that of family contacts of clinical cases. The Hopkins study was unique for its use of quantitative and statistical models to determine the risk from contact with carriers.[62] The number of expected cases due to exposure from carriers turned out to be significantly less than the number of actual cases reported. The authors concluded that the development of clinical diphtheria among those in contact with known carriers of virulent bacilli was at least an infrequent occurrence. They also showed that in cities where diphtheria was endemic, outbreaks were typically localized to a school, children's institution, or a neighborhood. In such instances, families with known carriers did indeed incur some risk of clinical infection, which though small, was still higher than in families where there were no carriers. Furthermore, they found no significant distinction between the risk of attack for those exposed to carriers with virulent or avirulent bacilli, or even for those where the virulence was not determined.[63]

The Hopkins study was an example of the increasingly more detailed epidemiological studies of carriers and their contacts in specific local contexts. The danger posed by carriers could only be seen and evaluated in such contexts where their movements could be tracked and any clinical cases in the vicinity could be linked back to them.

Conclusion

The carrier problem had initially been defined by bacteriologists in the 1890s. By the late 1920s, the carrier problem belonged to epidemiologists. Initially, health officers and physicians viewed the carrier problem as one in which laboratory scientists were pushing for restrictive practices and regulations for the control of diphtheria based on inconclusive data. The bacteriologists, in most instances, took a defensive position and attempted to clarify again and again the status of the laboratory data in relation to the diagnosis and control of diphtheria carriers. Yet they could not escape the fact that every study determined only the percentages of carriers in a given population (as von Sholly had noted).

Carriers existed, bacteriologists argued, and could transmit diphtheria to others; furthermore, they were everywhere in most urban areas and could not be localized in overcrowded tenement districts or in certain classes or occupations. The danger carriers posed was defined by the virulence of the bacilli they harbored. But virulence tests were widely acknowledged to be impractical for application on the large scale required. William Park, writing in 1923, acknowledged the practical limits of the bacteriological evidence:

> The necessity for immediate information forces the laboratory to report the result of a smear from a throat culture as positive without the confirmation of a virulence test. This is making a more definite statement than our knowledge really permits. All we know is that diphtheria-like bacilli are present in the culture. These may not be virulent. Knowing this fact we hesitate to adopt drastic procedures and we should always be ready to make a virulence test in any case in which the persistence of diphtheria-like bacilli makes a real hardship for the patient. The more doubtful the clinical diagnosis in a case, the more likely will the bacilli present be diphtheria-like rather than true diphtheria bacilli.[64]

Park would later acknowledge that the laboratory conception of virulence did not sufficiently answer the question about the infective potential of carriers. Only more extensive and detailed epidemiological studies could determine the extent to which carriers were a significant source of diphtheria infection, though they were themselves significantly less infectious than active cases of diphtheria.[65]

Bacteriologists largely disavowed any commitment to a particular method of controlling carriers, declaring it to be an administrative, not a scientific problem. Public health officers recognized that though carriers could not be dealt with in the same drastic fashion as "inanimate and probably innocent fomes," they realized that a great many people would have to be policed.[66] Ideologically, some health officers supported such an active state role in controlling carriers, believing that "medically and biologically, the interests of the whole, that is, of the race, are greater than those of the individual parts."[67] Yet the institutional apparatus to accomplish this magnitude of surveillance was clearly beyond the economic means of most health departments.

The failure of virulence testing and isolation to control carriers obscured the critical issue: Why were there carriers? Diphtheria infection involved the bacteria and the susceptibility or immunity of the host. Another alternative to the solution of the carrier problem lay in exploring the mechanisms of immunity in the host.

Historians have questioned whether a focus on carriers might have led to public health programs aimed at diminishing individual susceptibility to diphtheria through improvements in nutrition or housing reform.[68] Such programs did not emerge for several reasons. First, the bacteriological definition of the carrier focused on the characteristics of the microbe not the host. If the bacilli were virulent, the carrier was dangerous. Epidemiologists redefined the carrier in terms of its infectivity. The carrier's ability to produce diphtheria proved not to be as great as bacteriologists had assumed. The epidemiological evidence then provided a way to limit the detection, isolation, and supervision of carriers. Public health practice was dependent both upon bacteriological *and* epidemiological evidence. A turn toward the susceptibility of the host would have required a decoupling of public health practice from the prevailing bacteriological conception of diphtheria without which the concept of carrier literally had no meaning.

Though it was recognized that the carrier was probably immune to diphtheria, there was no way to evaluate how this natural immunity could be boosted by changing environmental conditions.[69] An understanding of immunity that allowed a role for environmental factors did emerge from the study of diphtheria carriers both from epidemiology and the emerging arena of immunology, but the discovery that toxin-antitoxin could be used to produce an artificial immunity to diphtheria pushed such questions into the background during this period.

6: THE USE OF ACTIVE
IMMUNIZATION

The successful immunization of thousands of schoolchildren with neutral mixtures of diphtheria toxin-antitoxin is considered by all observers as the pinnacle of the diphtheria program. William Park has justly been regarded as the prime architect of this culminating effort to control diphtheria in New York City. Yet there has been little discussion of why Park saw active immunization as the answer to the problem of diphtheria control or of how he was able to implement the program.

Few other cities adopted immunization as quickly as New York City.[1] In other locations, more questions were raised about the scientific evidence supporting the procedure.[2] That such questions were not raised in New York City is puzzling at first glance, given the scrutiny and debate surrounding other aspects of the diphtheria-control program in the past. However, the lack of debate over the implementation of active immunization is partly because after the turn of the century, the laboratory-based sciences of bacteriology and immunology had in many respects proven their efficacy in defining and controlling infectious disease. The process of defining diphtheria as a distinct disease with a specific cure that could be controlled only through the laboratory was complete by the first decades of the twentieth century. In New York City in particular, the Research Laboratory of the Department of Health headed by William Park had, by 1900, an international reputation for its laboratory studies on diphtheria and other infectious diseases and for the application of this research to the control of disease in the city.

Park's biographer attributes his success in the use of active immunization to the fact that he realized the significance of research on immunity to diphtheria that was being done in Europe and the United States and that he was willing to apply this work broadly.[3] In particular, it is argued that Park simply recognized the significance of Behring's aborted experiments on active immunization with toxin-antitoxin in humans and of Béla Schick's invention of a test to determine immunity to diphtheria. He was able to see that these tools were the solution to the problem of immunizing an entire population. Such a view, I suggest, disguises the many factors that led to the successful use of active immunization in New York City.

Park's success is more complicated than a simple story of scientific progress or a story of one man's insights. We have to consider how he and his colleagues viewed the problem of diphtheria after the turn of the century and also the scientific, technical, and institutional issues they had to address. As with the introduction of both the bacteriological diagnostic program and diphtheria antitoxin, the problem of diphtheria had to be made visible, and a solution to the problem had to address the continuing conflicting interests of public health, private medicine, and parents. Just as the diphtheria diagnostic program and the use of antitoxin for treatment were not the simple discoveries of one of Paul de Kruif's celebrated microbe-hunters, neither was active immunization just the "right" solution to the problem of diphtheria. By the first decades of the twentieth century, Park and his laboratory were linked in a complex web of social interests that now included pharmaceutical companies; municipal, state, and private philanthropy; private physicians; public school officials; and parents. The right solution to the diphtheria problem would have to provide a means to negotiate the interests of each of these groups.

It is also worth noting here that diphtheria itself presented new challenges for public health workers after the turn of the century. Though the number of deaths it caused had declined, it remained a prevalent and deadly disease in New York City despite the concerted efforts of the health department. The presence of carriers; the growing population of susceptible children in the city, especially in the tenement districts; the persistent problems of unrecognized and unreported cases of the disease; physicians' continued reluctance to administer adequate doses of antitoxin at the first signs of the disease—all these factors made Park and his colleagues aware that they had reached the limit of what they could do to control diphtheria with the tools they had in hand.

For Park the most promising solution to the problem of diphtheria had always centered around extending the immunity produced by antitoxin. He was uniquely positioned to draw from the several arenas where research on antitoxin generated insights about how active immunization might be achieved. Park engaged in and directed laboratory studies of the properties of the diphtheria bacillus. As director of antitoxin production for the city, he supervised efforts to improve the quality of antitoxin produced in horses. He directed the clinical use of antitoxin in the city's hospitals and in those patients monitored by the health department. He participated in and directed epidemiological and laboratory studies on diphtheria carriers. As the author of a major textbook on pathogenic microorganisms, he read and synthesized the international research on all aspects of the disease.[4] Finally, his location in the New York City Health Department gave him access to populations of children for inclusion in limited control trials, which allowed him to establish an empirical base for active immunization before it was applied to a wider population.

At one level, Park's success lay in his thorough understanding of diphtheria and his integration of laboratory research with clinical and epidemiological data. At the next level, he was able to go outside of the barriers imposed by the health department to marshal funds from private sources to support his work when public funds were not available. These funds allowed him to develop the empirical base on which he could build a case for active immunization. The data he collected were critical to gaining the support of physicians. Once the data on active immunization were collected, they became the foundation for garnering public support as well.

Ultimately, the success of diphtheria control in New York City involved substantial technical expertise. Sites for experimentation on children were needed along with the apparatus to promote experimental results and to persuade the larger medical and lay community to accept each innovation. In the case of active immunization, Park provided the technical expertise, experimental data, and methods for the broadest possible application of new techniques. The promotional and persuasive aspects of diphtheria control were managed by Hermann Biggs until the early decades of the twentieth century. Though Biggs died before the work on active immunization was complete, he left a powerful legacy in the health department. His successors, most notably Shirley Wynne, carried his legacy forward in the campaign to immunize preschool children that is the subject of Chapter 7.

Active Immunization

Late in 1894, according to William Park, an epidemic of diphtheria at the Mount Vernon branch of the New York Infant Asylum was "stopped immediately" by giving each child in the institution an immunizing dose of antitoxin.[5] Park and Biggs were so convinced of the immunizing value of antitoxin by this episode that the health department began the general use of immunizing injections two months later. Park later claimed that 25,000 immunizing injections were given throughout the city during that year.[6] The resulting immunity, however, did not last longer than three weeks.

In 1899 Hermann Biggs summarized the results obtained from the use of antitoxin in the treatment and prevention of diphtheria in New York City from 1896 to 1899. From Biggs's perspective, the antidiphtheria programs of the health department were an unqualified success:

> The hope earlier aroused in the minds of a few of the most enthusiastic workers in bacteriology have been more than realized with respect to diphtheria, and their most extravagant predictions have been actually exceeded by the results. The control of the disease, one of the most dreaded and fatal diseases in modern times, has been rendered easily possible.[7]

Biggs asserted that the control of diphtheria had essentially been achieved. While he attributed the decline in the mortality from diphtheria in this period to the use of diphtheria antitoxin (from 2,870 deaths in 1894 to 923 deaths in 1898, with a reduction in the case fatality rate from 29.7% to 12.2%), he also documented a decline in the morbidity rate. The reduction in actual cases of diphtheria, he claimed, was due to the unique practice of the New York City Health Department in giving immunizing doses of diphtheria antitoxin to those exposed to active cases of the disease. Few other cities employed antitoxin as an immunizing agent as widely as New York City. In comparison to other great cities, such as London, Berlin, and Paris, Biggs claimed that the "diminution in the actual number of deaths and the diminution in the death-rate per 100,000 population since 1894 has been greater than in any other great city in the world."[8]

New York City was different from these other cities in many regards, according to Biggs. First, he claimed, antitoxin was used widely and early, having a direct bearing on reducing the number of severe cases of the disease. Second, the use of antitoxin as an immunizing agent by health department inspectors and private physicians had largely reduced the number of secondary cases. Third, the department had begun the medical inspection of children in schools and increased the inspection of cases in tenement house districts.

Biggs's assessment of the impact of immunizing doses of diphtheria antitoxin on the reduction in morbidity in the city was somewhat suspect. He claimed that the number of cases had dropped from 2,359 in 1894 to only 788 in 1898 due to this practice. He was convinced that this reduction was due to the use of immunizing doses of antitoxin.[9] It is very difficult to find evidence for Biggs's position, since only a fraction of the total number of cases of diphtheria were ever seen by health department inspectors and their use of antitoxin for immunization was not uniform. While Biggs found this evidence persuasive, he also admitted that diphtheria had a higher prevalence in New York City than in either London or Paris. This was due, he noted, "to the great density of population and the character of the housing of the tenement house population, which is more favorable to the extension of infectious diseases than any city in the world," rather than to any inadequacy in the health department's practices.[10]

The prevalence of diphtheria in the tenement districts was still a severe problem at the turn of the twentieth century, although this became difficult to track due to the change in the city limits. The population of the city had increased with the creation of the Greater City of New York, which incorporated the boroughs of Queens, the Bronx, and Brooklyn. Some confusion occurs in evaluating the data on mortality and morbidity after consolidation in 1898, because much of the later data reported by the health department refers to aggregate city data rather than solely to the borough of Manhattan. There is, however, some data on antitoxin immunizations available for New York City.

The practice of the health department, beginning in 1896, was to give immunizing doses of antitoxin to all children, and sometimes to adults exposed to the disease, if consent could be obtained. After the injections were given, no attempt was made to separate the case of diphtheria from the family. By 1899 over 5,000 persons in 2,000 families had been immunized in the tenement districts.[11] Between 1902 and 1904 another 40,935 immunizing doses had been given by health department inspectors. A much smaller number (5,968) had been given to private patients by physicians between 1899 and 1904, according to health department reports.[12] Later the practice was extended from families where a case occurred to all children in families on the same floor of a building, again if parental consent could be obtained. The data presented do not indicate how many times this consent was refused or to what extent inspectors were able to immunize all the children in a building. Health department records showed that less than one percent of children immunized contracted the disease. But it is not clear, other than from the short trial reported by Biggs, if any control studies were conducted. Billings reported that another 23,042 persons had been immunized in institutions in the city.[13]

Such data suggest that immunizing doses of diphtheria antitoxin had some effect on the number of cases of diphtheria in the city. This was only true to a limited extent, given that fewer secondary cases developed from primary cases. However, there is no way to determine how thoroughly this work was carried out. Given the prevalence of diphtheria in New York City and the number of carriers, such short-term immunization was effective only if applied in a rigorous and systematic fashion. From reports of physicians working as medical inspectors in the health department at the time, there is much indication that this was not the case. By 1905, Park and his colleagues found that the number of cases of diphtheria in the city had not substantially decreased.

Immunization with prophylactic injections of antitoxin, while widely practiced in the United States after 1900, served only to control outbreaks of diphtheria when all contacts and carriers could be identified. It was therefore quite effective in institutions and often useful in controlling epidemics in sparsely populated rural communities where cases occurred in isolated homes. The effects of immunizing doses in controlling epidemics in such settings were quite dramatic. As one physician described in 1907,

It has seemed to me that antitoxin was a good thing to use liberally whenever diphtheria existed. In one county I rode twenty-seven miles, and in that time visited sixteen families and found six diphtheria corpses. Every child along the road was immu-

nized, and that was the end of the outbreak. If we had not acted upon the assumption of the possibility of the infection of other children, we would not have obtained such good results. We did not inquire as to whether the children were infected with virulent or non-virulent diphtheria bacilli. I believe this is good practice.[14]

Immunization was only useful when used on those directly exposed to diphtheria. Given the number of cases caused by carriers, there was no possibility of eliminating diphtheria by the use of immunizing doses of antitoxin. Park began exploring how immunization with antitoxin could be extended to protect persons who had no direct exposure to diphtheria. In 1922, he described his line of thinking:

The use of antitoxin in treatment was of course far more efficient because it could be used in every case as soon as it was discovered, but antitoxin for prevention could only be used where the danger was apparent from contact. Through the use of antitoxin the deaths have been cut down to about one-seventh of what we estimate they would otherwise be. The fact that the improvement which continued for many years after the introduction of antitoxin had ceased and that in fact a few years ago diphtheria began to increase slowly in the United States, made us appreciate that we had reached about the limit of what we could do with the old measures. It made us think seriously of using active immunization. This was the hope of rendering the population permanently immune rather than of waiting for cases to develop and trying to cure them and to prevent the spread of further contagion.[15]

The increasing incidence of diphtheria, the limitations of antitoxin, and the failure to control carriers convinced Park that some sort of active immunization was the solution to diphtheria control. However, Park fails to mention in this retrospective account two other factors that influenced his focus on active immunization before 1920: the inadequate work done by health department inspectors and the failure to control diphtheria in the tenement districts.

As mentioned above, Biggs and Park tended to downplay the problems in the application of existing diphtheria-control efforts in the health department. Perhaps they were conscious of needing to tread lightly on the problems that plagued the department in their public statements to avoid offending the city's political leaders. However, there is evidence from the published and private reflections of some of the physicians who worked in the department at this time that much of the work was done haphazardly. Few of the physicians employed to visit homes and enforce infectious disease regulations did so in a rigorous way. S. Josephine Baker, the physician who led the city's first Bureau of Child Hygiene and who worked closely with Park in the first trials on active immunizations of children, wrote in 1902, "Inspecting school children was my first assignment and it seemed to me a pathetic farce. . . . The only thing to recommend the whole dismal business was that it did, in a futilely primitive fashion, recognize that something might conceivably be done about controlling contagious diseases in school children."[16] Baker's comments and those of others on the work of the health department inspectors, of which Park was of course aware, suggest that he also might have considered the need to have a solution to the diphtheria problem that would minimize the need for oversight by health department inspectors.[17]

The continuing failure to control diphtheria and other infectious diseases in the tenement districts among the immigrant population also must have bothered Park. According to health department records during the first decades of the twentieth century, death rates from diphtheria within the immigrant population were high.[18] As birth rates in this population increased, so did the incidence of diphtheria and other infectious diseases. Among the highest rates were those in the Italian population. The death rate from diphtheria among Italian children was three times that of all other nationalities combined, though Jewish children had a death rate of almost twice that of the average for the entire city.[19]

As historians and demographers of this period have noted, it is difficult to get an adequate assessment of all the factors that contributed to the high mortality rates among the immigrant population. Much of the available mortality data is aggregated across the entire city of New York, and morbidity data for specific infectious diseases is difficult to find. Observers from this period—physicians, social workers, and health department workers—routinely attributed the high mortality rates of immigrants both to the conditions of the overcrowded neighborhoods in which they lived and to the distinctive cultural habits of each group.

Sympathetic observers attempted to provide more nuanced and complex assessments. Italian physician Antonio Stella, for example, conducted a house-to-house survey of several "typical" Italian blocks in the lower part of Manhattan in 1908.[20] Stella's survey is important because it is one of the few that collected statistics by age group for specific nationalities and specific diseases. The blocks selected were largely inhabited by Italians interspersed with Jews and African Americans. The disparities in the death rates between these groups was illuminating. On certain blocks in the Italian section, the death rate from diphtheria was among the highest in the city, while in other blocks less than half a mile away, in the Jewish quarter, the death rate was only half as great despite the similarity in conditions.[21] Stella commented, "In the latter [Jewish] district there is a greater population, the tenement houses are taller, and the general sanitary conditions are worse."[22] Stella attributed the disparity, not to the hygienic practices of the Jews, but rather to their "racial resistance" to disease that derived, he argued, from their long residence in cities in both Europe and America.

The situation for Italians in this part of the city was dire. Italian children also had higher rates of bronchopneumonia, measles, scarlet fever, whooping cough, and diarrheal disease. The Italians also had the largest number of children under five years in the sections Stella surveyed. Though contemporary observers have attributed the high child mortality to the effects of differences in breast-feeding and other nutritional practices, Stella concluded that the high mortality among Italians was a direct consequence of the conditions in which families lived in the tenement districts and the economic hardships they endured.[23]

Severe overcrowding was the norm, with many families occupying only one room, and almost half occupying only two rooms. Forty percent of the families took in lodgers since rent took as much as 33 percent of family incomes. Stella claimed that the level of overcrowding among Italians was worse than that among the other groups surveyed. He concluded that the overcrowding, along with working conditions both inside and outside the homes and the economic hardship that forced many Italian mothers and children to work in sweatshops explained the high rates of diseases like diphtheria and tuberculosis.[24]

Other observers confirmed the conditions that contributed to high rates of disease in the immigrant population. Economic pressures were such that even sick and convalescing children had to work as soon as they were able. Physicians were deeply troubled by the effects of these conditions upon children. Writing in 1905, Annie S. Daniel commented, "The other day a girl of eight years was dismissed from the diphtheria hospital after a severe attack of the disease. Almost immediately she was working at women's collars, although she was scarcely able to walk across the room alone."[25]

The presence of sweatshop work also undermined contagious disease control since medical inspectors were often barred from apartments and notices removed from premises in order that work not be interrupted. These practices led to many cases of diphtheria going unreported and prevented health department inspectors from administering antitoxin when it was most needed.

As John Duffy has discussed, this period also saw a return to the nineteenth century practice among immigrants of avoiding hospitals.[26] With much cause, there also continued to be a great deal of resistance and fear among the poor to having their children removed to contagious disease hospitals. In part this attitude reflected the way poor families were treated when their children were taken away. Parents were often not informed of the address or phone number of the hospital where their children were taken, nor were they given updates on their sick child's condition. When the reform administration of Seth Low took over the city in 1902, the new health commissioner, Ernest Lederle, acknowledged that these problems contributed to the spread of contagious diseases among the poor.[27] The health department efforts were also hampered by inefficiencies and personnel problems among its own ranks and the shifting political situation in the city between reform and Tammany Hall administrations.[28] Park's laboratory work, in particular, suffered during this period due to the constant changes in the funding for the laboratory's work . In 1903 the sale of diphtheria antitoxin to vendors outside of the health department was banned under pressure from physicians and commercial pharmaceutical companies. This funding had provided the bulk of support for Park's laboratory.[29]

It is clear from this brief discussion that during the period when Park began to consider the use of active immunization to control diphtheria, the health department and social reform groups were giving a great deal of attention to the ways in which the economic, cultural, and physical conditions within immigrant populations in the tenement districts contributed to the prevalence of contagious disease. Yet while efforts to control tuberculosis continued to be linked to discussions of the need to provide better housing and sanitoria for the poor, such issues were largely considered to be secondary factors in the control of diphtheria. It was possible for environmental factors to be downplayed with respect to diphtheria control because the use of antitoxin for passive immunization gave strong impetus to the view that a specific method for prevention could be found. Such a view was widespread in medical circles at this point. During the Eleventh Congress for Hygiene and Demography held in Brussels in 1903, for example, reports from France and other countries indicated that tests on the use of antitoxin for passive immunization were being conducted quite widely in orphan asylums, schools, and foundling homes.[30]

Experimental Research on Active Immunization

Experimental work on the immunizing effects of antitoxin largely built upon Paul Ehrlich's 1891 and 1892 work on the immunity produced in mice that were passively immunized with antitoxin.[31] Early on it was noted that injections of toxin that had been almost neutralized by antitoxin were capable of stimulating the production of antitoxin in animals. As Park described, this fact was only discovered accidentally during experiments on guinea pigs and horses.[32] In 1895, Wernicke experimented with the transfer of immunity in guinea pigs. He demonstrated that when guinea pigs were originally treated with toxin, then with antitoxin, and then again with toxin, the immunity passed on to their offspring was extended to almost three months.[33]

Park had independently confirmed the observation that toxin-antitoxin mixtures injected into horses would bring about the development of antitoxin.[34] In the winter of 1896, he followed up on Wernicke's work, using overneutralized toxin in starting the immunization of horses used for the production of diphtheria antitoxin. Four years later Theobald Smith at Harvard extended these experiments, using neutralized mixtures of toxin-antitoxin. "Animals which have passed through a severe disease due to toxin alone and manifested by a loss in weight, fever, extensive local necrosis and ulceration," he found, "transmit no immunity to their offspring."[35] While this result was not surprising, the finding that "a single injection of a toxin-antitoxin mixture which produced no local lesions so far as could be seen or felt and no distinct loss in weight, nor any general symptoms, induces an active immunity which persists for *several years*" (emphasis mine) had startling implications for the control of diphtheria. Smith concluded in this paper that if it were possible to contain the possible negative effects of the toxin, such neutralized mixtures "would be of great value to substitute for a *passive immunity* in exposed children an *active immunity* extending over a considerable period, provided such immunity is attainable easily and without any more difficulties than in the guinea pig" (emphasis mine).[36] Smith's straightforward suggestion on how the immunity against diphtheria could be extended in humans demonstrated that children would not have to suffer multiple injections with antitoxin in order to be immunized against diphtheria for longer periods of time.

William Park was very excited by Smith's experiments and by his suggestion that neutral mixtures of toxin-antitoxin could be used to extend the immunity to diphtheria in humans for as long as two years. Given that he himself had demonstrated how such mixtures were used to speed up the process of immunization in horses, the possibility of the application of such methods to humans seemed to him quite feasible.[37]

Six years later, in 1913, Emil Behring made the first injections of toxin-antitoxin mixtures in humans while working out a theory that claimed the origins of antitoxin in blood serum was a result of insufficient toxin in an organism.[38] This work was interrupted by the First World War, leaving a number of important questions about his experiment unanswered.[39]

There were a number of obstacles to be addressed before toxin-antitoxin could be used extensively in humans. Park's work on horses that were used to produce antitoxin indicated that it took a number of weeks after the injections of the neutral mixtures before the antitoxin developed in those animals not possessing it. Therefore, toxin-antitoxin would be of little use in persons in imminent danger of infection. In practical terms, this meant that health inspectors would have to make several visits to inject those who were in close contact with diphtheria cases. Smith's work also showed that the immunity produced from neutral mixtures lasted only two years. While this represented a significant increase over the immunity produced by antitoxin alone, epidemiological data showed that children were most vulnerable to diphtheria from ages one to five. Children would thus have to be reinjected every two years, an impossible task. It would also be difficult to determine if the immunization were successful without extensive trials. Some children would have to be injected and then have blood taken periodically to determine whether there was antitoxin present. They would then have to be compared with untreated children. In short, Park noted, "in the absence of any simple test for determining which individuals had natural antitoxin and which did not, we were under the necessity of injecting many unnecessarily, if active immunization were to be attempted."[40]

A further problem to the use of toxin-antitoxin mixtures for immunization was the severe reactions to antitoxin experienced by some individuals. The occurrence of rashes, swelling, fevers, and more severe, even potentially fatal responses had long been cited by critics of antitoxin. Park argued that these side effects had to do with the properties of horse serum. He conducted several experiments to improve the quality of antitoxin in order to eliminate such effects while preserving the greatest possible neutralizing action of the antitoxin.[41] Through this work he became aware of Schick's studies on allergic responses to antitoxin.

The Viennese physician Béla Schick, a long-time student of diphtheria, developed a test to provide the most important information needed to make active immunization in humans viable: the determination of immunity to diphtheria. Schick, along with all who had studied diphtheria, recognized that it was a disease of children, that adults were rarely susceptible to it. A number of studies confirmed that the answer to the riddle of adult immunity was that many individuals possessed antitoxin in their blood serum without having shown any symptoms of diphtheria during their lives. Schick had found that a large percentage of newborn babies, and up to 90 percent of adults, showed such antibodies. In Vienna he began to study methods for testing children and guinea pigs for the presence of antitoxin by injection of diphtheria toxin.[42] He also studied the side effects that arose from injections of antitoxin produced with horse serum.[43] This research, the definitive study of "serum sickness," illuminated the range of reactions that humans had to injections of antitoxin.[44]

In 1913 Schick published the results of experiments showing that diphtheria toxin exerted an irritating effect when injected into tissue, but that if blood passing through the tissue contained antitoxin, it effectively neutralized the toxin and no reaction occurred. A positive reaction to an injection of toxin indicated the absence of antitoxin or, in other words, a susceptibility to diphtheria. If there was no reaction to an injection of toxin, then antitoxin was present and the person was immune to diphtheria.[45] Schick mapped out for the first time, the age ranges which marked susceptibility to diphtheria. While 93 percent of newborns possessed antitoxin transmitted from their mothers, only 37 percent of children between two and five years old showed the presence of antitoxin. The percentages increased again from five years until adulthood.[46] Schick's test was a simple and direct method of determining immunity to diphtheria, and his results showed that during childhood and even in the period between two and five years, a considerable number of individuals were immune to diphtheria.

The implications of this result were not lost on Park. If Schick's test worked as claimed, then the diphtheria problem was not as large as it appeared to observers. The test could narrow the number of persons susceptible to diphtheria down to a finite number of people who could be fairly easily identified and rendered immune. Schick's work was widely publicized in Germany and Europe, although World War I prevented the use of the Schick test on a wide scale in Europe.

Active Immunization Begins in New York City

To-day in New York under the superb leadership of Dr. Park, and all over America, and in Germany, hundreds of thousands of babies and school-children are being ingeniously and safely turned into so many small factories for the making of antitoxin, so that they will never get diphtheria at all. Under the skins of these youngsters go wee doses of that terrible poison fatal to so many big dogs—but it is a poison fantastically changed so that it is harmless to a week-old baby!

There is every hope, if fathers and mothers can only be convinced and allow their children to undergo three small safe pricks of a syringe needle, that diphtheria will no longer be the murderer that it has been for ages.[47]

Park very quickly made use of the results of Behring's and Schick's work. Behring reported on the use of toxin-antitoxin in May 1913, and in the fall of the same year, Schick's paper was published. Within months, Park and his colleagues in New York began using the test along with toxin-antitoxin injections at the Willard Parker Hospital. In the first year about seven hundred patients admitted to the hospital, as well as staff and other adults, were given the Schick test and injected with toxin-antitoxin.[48] For two and a half years, Park and his colleague Abraham Zingher, a former student of Schick's, monitored the children in the study.[49] They found that children with a natural antitoxic immunity produced an increase in the amount of antitoxin after the injections, while those without any natural immunity, and thus probably susceptible to diphtheria, gave entirely different results. After about six weeks, 25 to 30 percent of the children without a preexisting natural immunity showed the presence of antitoxin. Six to eighteen months later a much larger proportion of them had a negative Schick reaction and thus were found to have finally become immune.[50] Park and Zingher initially viewed the fact that it took six to eighteen months for the immunity to develop as a late response. Their expectation from animal studies was that immunity would be developed more quickly, though they had yet to establish the precise dosage of toxin-antitoxin needed to provide immunity.

This first set of studies revealed that the toxin-antitoxin injections provided protection from diphtheria in school-age children lasting over a year and that no case of diphtheria occurred among those who had a negative Schick test.[51] Furthermore, local and general reactions, which occurred in about half of the injected individuals, were more frequent in older than in younger children and were usually not serious or long-lasting. At the end of this first study, Park and Zingher were prepared to recommend that active immunization be more widely employed by the medical profession, particularly on children in schools and orphan asylums; mothers and infants in lying-in places; and patients, physicians, and ward workers in contagious disease hospitals. They developed a set of guidelines outlining procedures for giving injections, determining the standard dosage of toxin-antitoxin, interpreting the tests, and monitoring pseudoreactions. They also offered to supply the reports of their own work as well as the toxin and antitoxin mixture to any physician upon request.

Park and Zingher's records confirmed Schick's results on the relationship between age and immunity to diphtheria. Nearly 50 percent of children between the ages of five and fifteen were immune to diphtheria. While about half of those between four and six years were immune, in younger children almost 60 percent were immune, compared to about 70 percent among adults. With these results in hand, Park and Zingher spoke about the Schick test and active immunization in forums all around New York City and at national meetings of the major medical groups. Park even took two students around with him to provide live demonstrations to physicians on the proper technique for reading the results of the Schick test.[52]

Park was called upon to teach physicians how to read the results of the Schick test on a number of occasions. The key was to distinguish the true reaction from a pseudoreaction. The true reaction was characterized by "a circumscribed area of redness and a slight infiltration" from one to two centimeters in diameter. This mark persisted for seven to ten days, and on fading showed some scaling and a persistent brownish pigmentation. The pseudoreaction appeared earlier, was less visible on the arm, and left only a faintly pigmented area after three to four days.[53] The true reaction took about thirty-six hours to come on, the pseudo, twenty-four hours.

While Park and Zingher always claimed that the Schick test was simple to perform and easy to administer (though with care), physicians seemed a bit unsure about the interpretation of the test. Park was repeatedly questioned about the timing of the appearance of the true reaction versus the pseudoreaction, the correct way to read the changes in skin pigmentation, the possibility of mixed reactions, the strength of the toxin used, and the duration of the immunity produced.[54]

Within three years, Park and Zingher had used the Schick test and toxin-antitoxin on some twelve thousand healthy children in different New York City institutions and in some fifteen hundred convalescing cases of diphtheria and scarlet fever at the Willard Parker Hospital. Zingher had designed a simple kit for mixing and dispensing the toxin and antitoxin in the appropriate dosage that allowed for efficient administration of the Schick test. They now had results that showed that the test allowed them to dispense with active immunization in about 70 percent of all cases.[55]

Though the test was a very reliable measure of the presence or absence of antitoxic immunity to diphtheria, they had also learned three important lessons. First, the toxin had to be of the appropriate strength; otherwise, no positive reaction would develop in those susceptible to diphtheria. With suitable toxin, Park claimed that the test gave "absolute security," that the person could not become infected with diphtheria. Second, pseudoreactions could also cause false positive tests. Park and Zingher therefore recommended the use of a control test.[56] Third, the test could not be read too early. Through trial and error Park and Zingher found that the reaction should be read at the end of twenty-four, forty-eight, and seventy-two hours.[57] Later data showed that the test could be read at different intervals, depending on the population being tested.

These early studies proved that the Schick test could be used to supplement the bacteriological and clinical diagnosis of diphtheria.[58] A positive Schick reaction supported a positive diagnosis of diphtheria when the clinical or bacteriological diagnosis was doubtful. In the Schick test, Park had a tool that eliminated much of the remaining uncertainty about the bacteriological diagnosis of diphtheria and the mechanism of immunity in the disease. Within two years, Park and Zingher began to urge that active immunization was the most important tool in the control of diphtheria. For individuals, Zingher argued, the assurance of immunity that active immunization carried would save "many an hour of worry over suspicious sore throats even after a bacteriological exam." For the community, he continued, the Schick test and active immunization had made the detection and elimination of diphtheria carriers unnecessary.[59]

Active immunization, however, was only possible with the Schick test. The test made it possible for the physician, or anyone, to see who was susceptible to diphtheria. In short, because the test result was visible on the skin, diphtheria carriers were visible as well. Since carriers were the key to the problem of diphtheria, from the perspective of the health department, the use of the Schick test would be crucial to mobilizing support for active immunization.

Following the positive results obtained at the Willard Parker Hospital, Park, Zingher, and Mae C. Schroeder extended their studies of the Schick test and toxin-antitoxin to a number of institutions in the city.[60] They began testing children at six-month intervals at various public-supported children's institutions. They sought out these institutions because children were held in them for a number of years, allowing monitoring over a fixed period of time, which was needed to establish the duration of immunity conferred by toxin-antitoxin injections.

While their reports on using the Schick test and toxin-antitoxin at the Willard Parker Hospital, orphan asylums, lying-in homes, foundling homes, and insane asylums are extensive, there is little evidence to suggest that any attempt was made to educate this population or their guardians about the procedure. In essence these children and adults were captive subjects for the studies being conducted by the health department. They were chosen, as Schroeder noted, because they were in institutions where they were likely to be held for a number of years, so the issue of parental consent was moot.[61] When this work was extended to children in public and parochial schools, however, the issue of parental consent could not be circumvented.

With the results of these studies in hand, Park approached the Board of Health in 1916 for funding to begin an active campaign in diphtheria preventive work in the public and parochial schools of the city. Denied funding for the project by the Board of Estimate, Park turned to private sources and received funding from the American Red Cross.[62] He obtained permission from the Department of Education to carry out the Schick test and the necessary immunization in about 250 schools. Little headway was made, however, because of the lack of necessary personnel due to the war.

School Work

With funds from the Manhattan chapter of the American Red Cross, Park was able to expand the active immunization work. By 1921, fifty-two thousand children in forty-four schools in Manhattan and the Bronx were given the

Schick test, retested, and those with positive reactions given injections of toxin-antitoxin.[63] Zingher and Schroeder had taken extensive measures to gain the consent of a number of people in order to carry out the diphtheria immunization work. First, permission had to be obtained from principals; then conferences were held with teachers, followed by the distribution of consent forms to parents of every child in the school. Teachers then prepared lists of those children who presented signed consent forms. Only these children were tested and given injections. "On the day of test," Zingher wrote, "the children were brought to a classroom where the physicians and nurses were doing the work. By good cooperation on the part of the principal and teachers, we were able to apply the Schick test and the control test to as many as 500 or 600 children each hour."[64] Thousands of children in a single school were tested this way. Certificates were presented to the children who participated that indicated either a natural immunity or immunity achieved after toxin-antitoxin injections.

The health department investigators realized that the support of principals, teachers, and parents was critical to their work. Their data had already shown that the group most susceptible to diphtheria were preschool children, yet since these children were scattered throughout the city in individual homes, they could not be easily reached. By giving immunizations to school-age children first, they reasoned that they could prove that the injections were safe and could simultaneously distribute educational material to parents through the children. Immunizing school-age children was also a strategy for preventing the spread of diphtheria to younger children in their families at home. Park and his colleagues claimed to have received an enthusiastic response to their efforts from teachers and school officials. Parental consent, however, was more difficult to obtain. "We are fortunate if consents are obtained from one-fourth of the parents," Zingher wrote.[65] The successful control of diphtheria by active immunization ultimately depended on their ability to educate and convince parents and others responsible for the welfare of children of the value of the procedure.

Since permission to test school children was granted in a reasonable number of instances, the studies continued. The numbers of children tested strongly suggest that Park and his colleagues had overcome a great deal of whatever resistance or concern parents may have had about the Schick test and the toxin-antitoxin injections, yet there was not universal participation in their program. One study of 192 families served in private practice showed that slightly over 15 percent refused to cooperate in the Schick test and toxin-

antitoxin prevention. There was no indication as to the nature of their refusal.[66] There may have been concerns about the safety of the procedure and
the duration of the protection toxin-antitoxin conferred.

Park and his colleagues promoted toxin-antitoxin as providing nearly absolute immunity to diphtheria, yet not all were convinced. At a symposium on
new methods for diphtheria prevention held at the New York Academy of
Medicine in 1922, Dr. Louis Harris warned against overstating the case: "We
should be guarded in our promises of ability to always confer immunity. We
should not instill the thought that the Schick test and toxin-antitoxin immunization offer a sure protection against diphtheria."[67] Harris reminded the
audience that the Schick test was not foolproof, and that if not carefully administered and evaluated, it could lead to erroneous conclusions. That same
year the annual report of the health department contained similar cautionary warnings about overestimating the duration of immunity conferred by
toxin-antitoxin.[68]

By 1926, however, Park was not very willing to admit to any limitations in
the duration of this immunity based on his seven years of study. His view of
the permanence of the immunity produced by active immunization was subsequently confirmed by many researchers across the United States and
abroad. However, it is important to remember that Park had the results from
one of the largest populations of immunized children anywhere. Only data
collected by the U. S. military raised some questions about the durability of
this immunity, and its data were difficult to compare with Park's. His belief
in the durability of immunity to diphtheria was also bolstered by personal
experience. In 1897 he had swallowed diphtheria bacilli and did not get diphtheria, whereas his colleague Anna Williams had a similar accident and had
come down with the disease. Williams's susceptibility to diphtheria was later
confirmed by a positive Schick test.[69]

The data from the schools provided some important epidemiological information about the relationship between density of population and the
number of children found to be susceptible to diphtheria.[70] Children from
the homes of the more well-to-do had a much higher percentage of positive
Schick reactions (indicating no immunity to diphtheria) than those from the
homes of the poorer classes living in crowded neighborhoods. When the
data was examined by race and ethnicity, they found that African-American
children living in congested neighborhoods also showed a high proportion
of positive Schick reactions.[71] Conversely, Italian children living in the
crowded East Harlem section of the city had the lowest percentage of posi-

tive reactions; children of Bohemian and Irish descent also fared well. Jewish children, however, varied widely in their reactions. Those living in congested areas showed a low percentage of positive reactions, while those living in neighborhoods on the Upper West Side were among those with the highest percentage of positive reactions.[72] These variations seem to indicate that residents of wealthier areas had less natural immunity to diphtheria. This result indicated that there was some other factor besides age at work in the production of antitoxic immunity to diphtheria.

Zingher speculated from this data that the relative segregation of well-to-do children in private homes meant that in early childhood they had little exposure to diphtheria and therefore developed little natural immunity to the disease. Repeated exposure to cases of diphtheria in the congested districts of the city allowed for the gradual development of a natural immunity.[73] In some ways this discovery confirmed the views expressed by Abraham Jacobi and clinicians of an earlier generation that contact with people living in the tenement districts exposed individuals to diphtheria, yet that paradoxically it also protected them. Their environmental explanation was incomplete, but Zingher's data did confirm that Jacobi was right to focus on the role of mild cases in diphtheria.

The results of these Schick tests from such a large number of school-age children explained for the first time some of the more puzzling aspects about the distribution of diphtheria in the city. Clinicians had long noted that in outbreaks of the disease children in one family were often struck, while children living in families in similar circumstances and located in close proximity to those with the disease did not always develop it. The Schick test results showed that there was a marked tendency for all children in one family to have the same Schick reaction, and that when there were both positive and negative reactors present, it was the youngest member of the family who was susceptible. This finding suggested that there was also a hereditary factor in the development of diphtheria immunity. Many observers later confirmed this point, which, as a rough general rule, became known as "Zingher's Law."

Zingher's results further indicated that race was a factor in the susceptibility to diphtheria. This was the first time that race had been made an explicit factor in discussions of diphtheria in New York City. Certainly in the period before the turn of the century, some attention was paid to ethnic disparities in incidence of the disease; yet none of the health department data had shown that race per se was a factor in diphtheria. This was partly because the case report of communicable diseases in New York City did not include an item on the race or color of the patient.[74] Racial or ethnic factors could be assessed by means of census data, which contained racial and ethnic characteristics of the population in each sanitary district. The high rates of diphtheria that had been seen among African Americans had long been attributed to the fact that they lived in poor, overcrowded districts of the city. The toxin-antitoxin immunization program provided the first direct information of the differences in immunity to diphtheria between whites, Blacks, and other ethnic groups. Haven Emerson's retrospective evaluation of the health department data from 1921 through 1925 showed that the Black and Irish population of New York City had the highest morbidity and mortality from diphtheria in the city.[75] Emerson concluded that race was the only factor that explained the differences between native-born whites and these two groups because he assumed that there were "uniform conditions of contact and exposure to diphtheria among all groups in the city."[76] The issue of racial differences in diphtheria incidence and immunity did not become a focus of the health department's attention in New York City; however, questions about racial differences in diphtheria were of great interest to public health officials in Baltimore and other parts of the South.[77] Park and Zingher's data brought international attention to the study of racial and hereditary factors in diphtheria immunity.

Convinced by their work with school-age children in 1918, Zingher published results of his observations on the duration of maternal immunity to diphtheria in young infants and active immunization with toxin-antitoxin in mothers and infants at the New York Foundling Asylum.[78] He found the largest proportion of positive Schick reactions in infants between the ages of six months and four years. The positive reactions diminished rapidly after the fourth year of life. In addition, Zingher found that infants suffered little or no constitutional or local reactions to toxin-antitoxin injections. However, some of the mothers, especially those showing pseudoreactions, had rather severe local and constitutional symptoms "consisting of redness, swelling and tenderness of the arm, headache and slight febrile disturbance, lasting about two days."[79] The work of immunizing infants and preschool children, however, proceeded slowly because Zingher and Baker found it difficult to persuade mothers that immunization was necessary.

By 1922 Park, Zingher, and Schroeder had conducted extensive studies among school-age children and some limited studies of infants and preschool children. They had demonstrated that the Schick test was a relatively safe and reliable measurement of antitoxic immunity. They had delineated the duration of the immunity conferred by toxin-antitoxin injections and the range and extent of problems associated with the Schick test. Much of the work conducted after this point on larger groups of children simply confirmed the results of these first studies.

The evidence suggested to them that active immunization against diphtheria was the most effective prophylaxis available to control the disease. While further improvements in the method of immunization were possible, they argued, there was no reason to delay the employment of active immunization against diphtheria. In sum, they felt that if their procedures were followed, it would be possible to create a diphtheria-immune population within a few years.

While Park, Zingher, and Schroeder were justifiably pleased with what they had accomplished in New York City, it is also worth noting that they were, in one respect, incredibly lucky. Working with small teams of nurses and physicians, they had successfully tested and injected large numbers of children with toxin-antitoxin without serious mishaps. Other cities were not as lucky. In Dallas, for example, a mix-up in the preparation of toxin-antitoxin that was not properly neutralized led to the deaths of several children. In Boston, toxin-antitoxin that had been frozen and then thawed was injected into a number of children. They developed severe cases of diphtheria, though there were no deaths.[80] Park's group had control over a number of factors that caused problems in other cities. They produced their own toxin-antitoxin and perfected the use of the Schick test. Due to financial and personnel constraints in the health department, Park, Zingher, and Schroeder had personally conducted many of these Schick tests and immunizations, perhaps minimizing the possibility for errors and mishaps that occurred with more inexperienced workers.[81]

I do not mean to suggest, however, that Park and his associates, because of their celebrated expertise, stood apart from the many practical dangers and ethical complexities associated with bacteriological work and the use of human subjects in this period. Zingher's untimely death is a solemn reminder of these issues. On June 6, 1927, the wife of Dr. Abraham Zingher found him asphyxiated in his office at the City Research Laboratory. According to the report in the New York Times, a gas tube had become detached from a Bunsen burner on the table where he was working on improving the Dick treatment for immunization against scarlet fever. Zingher was forty-two years old at the time of his death. The Times noted that he had conducted many experiments on immunization techniques against diphtheria, scarlet fever, and measles on children at the Israel Orphan Asylum. His wife confirmed that he had also successfully tested serums on their four young children.[82]

In an editorial about Zingher's death, the *Times* lauded him as a hero to scientific experimentation. "Heroes are to be found in all modern crusades," they wrote, "none braver than some of the quiet, dogged workers of the laboratories. . . . Other similar work needs to be done with scarlet fever and measles and other diseases by the fighting chemists who, like Dr. ZINGHER, risk their lives in the battle."[83] The editors used the occasion of Zingher's death to speak against opposition to experimentation in medical research. By experimentation they meant the use of human subjects to test new treatments and cures for disease. The editorial mentioned an army private serving in the Philippines who had volunteered to be inoculated with the germs of dengue fever in order to test a new remedy. They also noted that the opposition to diphtheria immunization due to the severe reactions following the injections had subsided after the materials and technique had been perfected. In lauding Zingher's work, however, the editors failed to mention the ways in which children had been used in the process of perfecting these same techniques.

While the evidence to date is limited on the harmful effects of the active immunization trials on institutionalized and public school children, it is clear that Park's use of children to test other vaccines came under increasing scrutiny in the 1930s.[84] There is evidence that certain practices had become routine in Park's laboratory in the years from 1910 into the 1930s. Among them were the use of children at the Willard Parker Hospital for numerous clinical trials on a variety of vaccines beginning with the work on diphtheria antitoxin in 1894. Second, there is evidence that workers in the lab, including Park, also tried out new vaccines on themselves and their families.[85] Zingher and the health department were publicly criticized for their methods of testing of sera for polio therapy in the 1916 epidemic.[86] The *Times's* silence on the use of children in the active immunization trials should not be seen as evidence that the use of children in the first tests of toxin-antitoxin was unproblematic or universally condoned.

Conclusion

The invention of diphtheria antitoxin by Behring in 1890 opened the modern period of experimental research on the mechanism of immunity in diph-

theria and led to the research that finally made active immunization of humans possible. Within months after the publication of Schick's work and
Behring's use of toxin-antitoxin on children, Park was convinced that the
major scientific questions about immunization had been solved, and he
moved rapidly to implement a program of immunization in New York City.
He was so quickly convinced because he had been working on the problem
of immunization both in horses, via the work on the production of antitoxin,
and in children, through the use of antitoxin for passive immunization in
New York City since 1894. The Schick test and toxin-antitoxin constituted an
oblique attack on diphtheria, given that, as Louis Harris noted, "the direct
attack of terminating the carrier state had failed."[87]

By the time active immunization was presented to the medical community in New York City, Park had a large database of empirical studies of its use
on children and adults in institutions around the city. Initially, physicians'
questions about the procedure were defused by the results of these studies
as well as by their familiarity, if not actual experience, in using antitoxin for
passive immunization. Park, Zingher, and Schroeder took every opportunity
to present their data to the medical community. By all accounts the impact
was positive. Comments such as these from Dr. Edward Bauer of Philadelphia support this claim: "When we hear of men testing 50,000 children, it
requires considerable courage to tell you we have personally tested 5,000
children. . . . Philadelphia does everything conservatively, but 'gets there' in
the end."[88] It is troubling that this data, however persuasive, had been gathered under conditions where institutionalized children and adults often had
few protectors.[89]

Yet Park was not unaware of the problems of convincing both parents and
physicians of the need to employ active immunization for all children.
Though some segments of the medical community had validated immunization, this did not mean physicians would simply begin to use it as a routine
part of their practice. As in the case of antitoxin, a public forum was needed
to persuade physicians and the public of the value of immunization.

Retrospectively, Park highlighted his interest in the immunizing value of
diphtheria antitoxin.[90] He had always been more fascinated by antitoxin's
immunizing effects than by its curative ones. For Park, active immunization
meant the complete control of diphtheria. This theme is at the center of his
accounts of the diphtheria program in New York City.

Antitoxin's failure to provide long-lasting immunity made active immunization the only solution to the diphtheria problem because it offered com-

plete control. But Park also recognized that such a solution had to minimize surveillance and intrusion in order to win public support. The Schick test, like Park's culture kit, was a portable tool that could be used outside of the laboratory—it too brought the laboratory to the streets. And like the culture kit, it made the problem of diphtheria, now defined in terms of how to protect the susceptible from the immune, visible to all.

7: THE DIPHTHERIA
PREVENTION COMMISSION

By 1921 Park and his colleagues at the New York City Health Department had successfully used the Schick test on more than fifty-two thousand schoolchildren in forty-four schools in Manhattan and Brooklyn. Children with positive reactions to the test were injected with toxin-antitoxin. By tracking these children, Park and his colleagues gathered important information about the duration of the immunity produced by toxin-antitoxin, the accuracy of the Schick test, and the need for retests. Although this research was successful in demonstrating the value of Schick tests and toxin-antitoxin immunizations, it only underscored the magnitude of the problem of immunizing all the susceptible children in New York City. While many of the school-age children could be immunized in school by health department physicians and nurses, the half-million children treated by private physicians could not be immunized this way. Most important, preschool children, those under five, who were the most susceptible to diphtheria, could not be reached. Those familiar with other public health projects among New York City's heterogeneous population realized that a major effort would be needed to convince parents to allow their infant children to be immunized by either public or private physicians.

Acknowledging the failure of all other methods thus far to control the number of cases of diphtheria in the city, the health department viewed the active immunization of the preschool population as critical. The funds needed to mount a campaign to reach this group, however, were not forthcoming from the Board of Estimate. The 1920s were years of fairly constant upheaval in the health department as Tammany Hall's control over the city's finances and the Mayor's Office wreaked havoc with both the personnel and the projects of the department. Park turned to private sources for support in his determination not only to control diphtheria, but also to eradicate it. In response to funding problems for the antidiphtheria project as well as to the more general problem of inadequate health care services for the poor, local health centers were established across the city. The Bellevue-Yorkville Health Demonstration Project, funded by a number of social welfare agencies and the health department, became the testing ground for the immunization of preschool children against diphtheria. The health department's collaboration with social welfare agencies in public health work was viewed skeptically by physicians. They saw in such projects the potential for the erosion of their client base and the specter of state medicine. Building on the methods found to be successful in the immunization of preschool children against diphtheria in the Bellevue-Yorkville district, the health department moved to extend the work throughout the city.

In 1929, with the data from the demonstration project in hand and an increasing incidence of diphtheria in the city, newly appointed Commissioner of Health Dr. Shirley Wynne marshaled an impressive group of philanthropists, social welfare agencies, and physicians to launch one of the first and most extensive public immunization campaigns against diphtheria. The goal of the campaign was to completely eliminate the disease in New York City within two years.

The Diphtheria Prevention Commission launched a massive public health campaign. Spearheaded by the commission, it used every media outlet from radio and newspapers to movies and billboards. Major institutions in the city were enlisted, such as the Catholic Church, the Metropolitan Life Insurance Company, and local gas and electric companies. Immunizations were made available to the public in schools, public beaches, and parks. Healthmobiles traveled through every neighborhood in the city. In the wake of this unprecedented attention, the image of diphtheria was transformed. Until this time, diphtheria had been a common childhood disease that, though it could be treated, still posed a serious threat to children's lives. During the campaign, however, diphtheria was portrayed as a disease whose very presence was a disgrace, a visible sign of parental neglect and medical indifference.

The campaign launched by the Diphtheria Prevention Commission came at the end of the Progressive Era. The role of the state in the protection of children's health, particularly poor children's, had come to the fore nationally with the formation of the Children's Bureau in 1912, and locally with the establishment of the New York City Bureau of Child Hygiene in the Department of Health in 1908 under the leadership of Dr. S. Josephine Baker. With the passage in 1921 of the Sheppard-Towner Act, which funded health services for mothers and children, the issue of the responsibility of the state for children's health was hotly contested. The response of physicians to the efforts of the New York City Health Department to control diphtheria reflected all the tensions generated by this issue.

The control of diphtheria in New York City had been a major project of the health department since 1893. The department had consistently taken the lead in introducing more scientific and effective methods to control the disease. Physicians had long challenged the role taken by the department in promoting methods to control diphtheria that relied heavily on bacteriological and now immunological research. Physicians' criticism continued after the introduction of methods of active immunization against the disease in the 1920s. Nationally, medical opposition to the state's role in preventive medicine was at a high point during this period. Locally, in New York City, with political control shifting back and forth between Tammany Hall and reformers in this period, the projects of the health department suffered. Privately funded clinics and social welfare agencies began to fill the gap. The lack of coordinated efforts meant that the health needs of the poor and working-class population were largely a hit-or-miss affair.

Thus the highly visible and focused health campaign against diphtheria filled the void created by the failure of organized medicine, the city, and private philanthropy to fully mobilize against diphtheria. The campaign displayed the Progressives' faith in science and their persistent concern with the relationship between science and social betterment. In this campaign, as in other Progressive Era reform programs related to children's health, maternal ignorance was blamed rather than the failure of private medicine or the state in providing preventive health care for all children.

The media was critical of the Diphtheria Prevention Commission's campaign to create a "diphtheria conscious" public in New York City. Although there were differences between the use of the media in this campaign and that of 1894 which introduced antitoxin, in both cases the press failed to take a critical approach toward uncertain scientific information. Like others in this era, the campaign was designed to promote the power of medical technology and the virtues of personal hygiene. It encouraged people to consult only medical specialists and warned of the dangers of ignoring medical advice.[1] The rhetoric of the campaign promoted the prevention of diphtheria as the result of a progressive science that had brought the disease to its proverbial knees. More important, this public campaign turned diphtheria control into a dramatic event that raised public expectations through rhetoric that transformed scientific progress into simple "common sense."

In 1929 the scientific, medical, and epidemiological questions about the treatment of diphtheria and the role of the carrier in transmitting the disease were overshadowed by the campaign against diphtheria in New York City. The lack of complete knowledge about diphtheria was seen by public health experts as less important than their belief that it could be controlled almost perfectly by active immunization.

Prologue to the Campaign: The Bellevue-Yorkville Health Demonstration

In 1921, Mr. George Bedinger of the American Red Cross agreed to support William Park's diphtheria immunization efforts. While acknowledging that public health was the job of public authorities in the long run, Bedinger noted that the Red Cross was interested in supporting projects that involved bringing together various agencies interested in public health projects in communities. Park's proposal for control of diphtheria seemed a worthy one to Bedinger.[2] At that time the American Red Cross established a health center in the East Harlem District of the city. This center, which opened in 1921, served an area of the city with a population of about one hundred thousand, largely Italian immigrants. It was the first such center that included both private social welfare and public health agencies.

Commissioner of Health S. S. Goldwater formulated a plan in 1914 to "test the value of local administration of the functions of the Health Department."[3] The need for a local focus for health department activities reflected the belief that "effective application of health programs, especially among the poor and the foreign-born, required an approach to the people on their own ground, in their neighborhood."[4] The East Harlem Health Center and the others that followed it were the outgrowth of Goldwater's policies. As George Rosen notes, this movement toward community health centers was a reflection of the complexity of the health problems and needs of immigrants living in deplorable conditions in the city and a realization that effective remedies required the coordination of efforts by public and private agencies.[5] Settlement workers, visiting nurses, and other observers had noted that people living in these communities often failed to use services located in strange and unfamiliar sections of the city for fear that they might encounter language and other barriers. The move toward community health centers also underscored the fact that "the problems for which poor people needed help were usually neither simple nor single and had no easy solutions."[6]

The health department initially established a few trial health centers on the lower east side of the city and in Queens. According to Winslow and Zimand, the success of these projects supported the belief that a "complete public health program necessitated exercise of all Health Department functions in a given neighborhood." The projects were then extended to other areas in the city.[7] A major impediment to the implementation of the health center program in the city was the lack of public funds to fully sustain the centers. Given the political situation in the city, it was unlikely that public funding alone could be relied upon, and thus private support for the program was sought.

The Bellevue-Yorkville Health Demonstration project opened in 1926 with the first clinic for preschool children. It was located in an area of the city bounded by 14th, 42nd, and 64th Streets and the East River, 4th Avenue, and 6th Avenue. The district contained a largely Irish and Italian population of about 174,000.[8] At the opening of the demonstration, sixty-five different public health and private welfare organizations were operating in the district, including the health department, the Henry Street Nursing Service, the Maternity Center Association, the Association for Improving the Condition of the Poor, and the Charity Organization Society. There were special clinics located in the district for treatment of tuberculosis, for dentistry, and for mental health, as well as a number of hospitals and dispensaries. Despite this plethora of institutions and services, the death rate for the district was 43 percent higher than that of the city as a whole.[9]

The Milbank Fund committed one million dollars to the Bellevue-Yorkville project. It was managed and operated by a Community Health

Council, which included the city's commissioner of health, members of the city's boards of health and education, and representatives of the various social welfare organizations that had long worked in the area. The expressed objectives of the project were to emphasize that the official responsibility for health work in the district rested with the New York City Department of Health and thus the project was to function as a supplement and aid to its work. The governing structure of the demonstration made the city's commissioner of health the chief executive of the project. The demonstration was thus carried out under the direct leadership of the health department, and most of its undertakings were projects of the department that were now coordinated with the work of the other agencies in the district. The projects undertaken in the Bellevue-Yorkville district by the health department were designed as prototypes. If they were successful in that district, they would be applied to the entire city.

The demonstration project was organized for work in three broad areas: preventive medical service through clinics, public health nursing, and health education. From the beginning, the health of infants and schoolchildren was a primary focus of the project. With a total of twenty-four elementary schools serving over ten thousand students, and with more than three thousand births per year, the district had a large population of children and infants.[10] The living conditions for children in the district were not good. During the period 1922–1926, before the demonstration project began, the infant mortality rate averaged 37 percent above the general rate in the city. For practically every cause of death, the Bellevue-Yorkville district mortality exceeded that of the city as a whole.

The data collected by the demonstration showed that the school-age population was indeed neglected. In a study of third graders, it was found that the overburdened school medical inspectors had ignored many of the physical conditions of the children. Problems resulting from poor nutrition and inadequate attention to tonsils and eyes were found to be widespread.[11] The central component of the health demonstration project's work included preschool clinic services, health education, medical inspection in the schools, pre- and postnatal care of infants, and educational programs to instruct mothers in modern child-care methods.

The preschool clinics in the Bellevue-Yorkville district grew out of the program begun earlier by the health department to address the needs of infants and children. In 1908 the department had established the Division of Child Hygiene under Dr. S. Josephine Baker. Four baby health stations under the direction of this division were located in the Bellevue-Yorkville district. These baby health stations were critical to providing care to infants and preschool children. The health department had long realized that the biggest impediment to public health work with this population was that there were no local or central locations where children could be treated. Visits into the homes of mothers of newborns and preschoolers by public health nurses were useful, but not the most effective or efficient way to reach this population.

Any effective diphtheria prevention program had to be able to reach the population of infants and preschool children as well as those of school age. Yet it was not easy to convince the Board of Estimate to provide funds for health projects for young children. As Baker noted, it was the "neglected, unhappy misfit stepchild, known as the 'preschool child,' who [was] the logical point of attack."[12] It was precisely this preschool child that was key to the solution of health problems in the schools with respect to diphtheria.

Early on, Park had enlisted the help of Baker's staff in reaching the preschool population. The issues to be addressed were different from those of school-age children in some respects. Public and parochial schoolchildren were a captive population. Once the Board of Education or the Catholic leadership was convinced of the value of immunization, principals and teachers were compelled to go along with the program. Teachers were urged to include lessons on the value of immunization as a part of the overall health education curriculum.

In encouraging the students to return permission forms for immunizations, the hope was that students would in turn educate their parents. It was recognized early in the initial immunization of schoolchildren that the education of parents was critical to the success of the program. Baker recalled in her memoir that mothers would be more likely to allow their babies to be immunized if their older children were successfully immunized:

> We knew we should have a fearful battle introducing mothers to let us test their babies, whereas permission to test older children was more easily gained. When a mother had once known her older child to come through such a test unharmed and when, after subsequent injections of toxin-antitoxin she was convinced that her child was protected against diphtheria, it was not too difficult to let her baby have the Schick test too.[13]

Authority figures, such as principals and teachers, served as intermediaries between the health department and parents. With the preschool and infant

population dispersed throughout a community with no institutional affilia-
tions, public health workers had to deal directly with the parents of these
children, most often the mothers. The attitudes of the mothers had to be directly confronted in order to successfully implement a diphtheria immunization program for infants and preschool children. The Bellevue-Yorkville
Health Demonstration project provided an opportunity for public health experts to determine the concerns of these mothers in order to devise effective
educational methods to address them.

Baker's comments point to a central issue that would be exploited in the
Diphtheria Prevention Campaign, the need to persuade the so-called fearful, recalcitrant mother, who was perceived as being unreasonably unwilling
to allow her infant child to be protected (in this case by immunization) by
health authorities. This construction of the fearful, recalcitrant mother was
pointedly directed at poor immigrant women living in the poorer districts in
the city, in populations that had been traditionally targeted for health department programs.

Having identified "fearful and recalcitrant mothers" as a problem, more
information was gathered to outline the extent of the problem. Through the
Bellevue-Yorkville project, data were collected that revealed the range of attitudes toward immunization in the district. Park's department assigned a
staff member to call on all the physicians in the district. Personal visits were
also made to school principals in the area. Meetings were held with social
workers and nurses as well. Still, by 1927, little information was available
about how many children had been immunized in the district.

Miss Margaret Newman, a graduate nurse, was engaged to investigate the
situation by visiting all families in selected blocks within reach of the Health
Center. She visited 235 families, going from house to house. According to her
report, a significant number of the children surveyed had not been immunized. Only nine of the families claimed to have heard about the toxin-
antitoxin inoculations. The reasons given by the respondents for not having
their children immunized were illuminating. Some claimed that they did
not believe immunizations were effective and feared their bad effects. Others stated that their private physician had not thought immunizations were
necessary for their children. A few claimed that their priest or some agency
that they dealt with had spoken against immunizations. Some of the parents
believed that it wasn't necessary to have their children immunized until they
entered school. A number of the respondents were working mothers who
either could not afford the immunizations or could not get time off from

work to take their children to the clinic or pay for a private physician. Some just claimed that it was too hot to bother with such things. In many cases Miss Newman had to return in the evening to talk with a father because the mother did not understand English or she put the responsibility for not having her children immunized on the father.

Nurse Newman also noted ethnic differences in the responses of the families surveyed. Among Italian families, she found that the mothers were often willing to have the children immunized, while fathers spoke against it. Some of the fathers noted their bad experiences with injections in the army or articles they had read in the papers arguing against immunization. One woman claimed that her two older children had been immunized four times by different agencies; therefore, she refused to have her three younger children immunized.

Among Irish families, which Newman described as being of "good standards, many of whom were born and brought up on the block," she noted a "pronounced hostility" to immunization. "The parents took the position that the children belong to them; that there was no law requiring toxin-antitoxin and until there was they are not going to have it done." Newman concluded that among families who had not had their children immunized, there was a great deal of deep-rooted prejudice against immunizations. She ascribed these feelings to ignorance and misapprehension. Furthermore, she found that the printed material available was of little use in overcoming the concerns these families expressed. Nurse Newman ascribed much of the antipathy toward immunization in the district to the "confusion among them, caused by the large number of health measures urged upon them from all directions."[14]

Nurse Newman's report typifies the evidence used to define the barriers to immunization in poor communities. These barriers were then strategically recast in terms that allowed interventions to be developed. First, the "fearful and recalcitrant" immigrant mother would be described in terms that represented her refusal of immunization as being the result of ignorance and misplaced fears of scientific authority. Second, immigrant parents' notion that they held the ultimate authority over their children would be cast against the need to subordinate such views to the greater good of the larger community.[15] And last, health authorities would avoid the very real problems of the fragmented health care in such communities by speaking in one unified voice in support of immunization.[16]

A somewhat different approach was taken with respect to the more privileged classes in the city, those, more specifically, who received their medical care from private physicians. As early as 1924, the health department began to solicit mothers directly, encouraging them to demand that physicians immunize their children. That these mothers were urged to take their children to private physicians rather than to the health department reflected the traditional strategy of the department with respect to private medicine and diphtheria control. The department maintained that the responsibility for protecting children from diphtheria was an integral part of the duties of mothers and private physicians. Though appearing to support private medicine, the appeal was designed to enlist this group of mothers to gain physicians' support for a health department project that had been received with skepticism by private medicine. The education efforts directed toward mothers had the effect of building an alliance between mothers and the health department against private medicine.

Unfortunately, the rhetoric of the campaign would more overtly target the so-called failures of mothers than the failings of private medicine in the immunization of preschool children. The health department's annual report reflected their view of this relationship:

> Under present conditions, two remedies would seem to promise the best results. The first is to change the attitude of the public, so that they will understand that it is not enough to surround maternity with such safeguards as are now increasingly utilized in pre-natal and puerperal supervision. They must be taught to demand of their physicians continuous and unremitting oversight to assure the utmost safety as to preventable diseases.[17]

The health department cast every death from diphtheria as a preventable death. Thus all deaths of children from the disease were an accusation against the community and as such were a public concern. The responsibility for the failure to have children immunized was placed directly on mothers. Attitudes toward mothers were little different from those expressed by health department officials and reformers with regard to other contemporary issues affecting the survival of children. Proponents of "scientific motherhood" emphasized that women needed expert advice in order to perform their duties as mothers successfully.[18] Language about the "neglect" of children by "ignorant" mothers was prevalent in a number of the child-saving campaigns of the period. The perceived failure of mothers to respond to the appeals of public health reformers laid the basis for greater professional intervention in a range of child-rearing and child-health practices, particularly with respect to preventable diseases.[19] Public health efforts in the control of disease in this period were conducted in the name of science. Ultimately, it was not the behavior of mothers that concerned health officers, but the implicit challenge to the authority of science that their resistance embodied.

Physicians' resistance to immunization was cast in similar terms. In mounting frustration, the health department took the position that it had done everything possible for the control of diphtheria. The only direct impediment to the complete control of the disease in the city, in its view, was the failure of physicians to immunize children. Between 1922 and 1929, the failure of private physicians to take up this work was a continuing point of contention. Their failure to immunize children was attributed to a range of issues. First of all, the familiar problems with the identification and treatment of diphtheria cases persisted through the 1920s. The health department claimed that physicians were too hesitant in their use of antitoxin when faced with parental objections. Furthermore, they were accused of delaying treatment with antitoxin either because they waited until the diagnosis was confirmed by a laboratory report or because they feared that their patients would suffer adverse reactions to it.[20] Since the extensive use of antitoxin had shown that it was relatively harmless when used in doubtful cases of diphtheria, the department had changed its recommendations and now encouraged physicians to use antitoxin as soon as possible in all such cases. Health department reports expressed a growing frustration with physicians' reluctance to immunize children. It had provided the medical community with numerous reports attesting to the safety of toxin-antitoxin along with the data collected by Park on thousands of schoolchildren who had been safely and successfully immunized against diphtheria in the city. But physicians were not convinced.

The health department began issuing circulars jointly signed by the county medical societies in its campaign to get physicians more involved in diphtheria prevention. The circulars emphasized that the control of diphtheria was in the hands of physicians.[21] The department was willing to provide physicians with placards for their offices and any other educational materials they needed in order to encourage parents to have their children immunized.

Despite the efforts of the health department to reach physicians and parents, the number of cases of diphtheria in the city remained high. Though the number of deaths due to diphtheria was declining during the 1920s, the health department viewed all deaths from diphtheria as preventable and expected the mortality to be much lower. This was especially true, since with the testing and immunization of 750,000 children in foundling homes and orphanages, the disease had been practically wiped out.

In 1928 when Dr. Shirley Wynne was appointed commissioner of health, he was extremely dissatisfied with the incidence and mortality from diphtheria in the city. The health department's campaign to enlist physicians in the control of the disease up to this point had obviously failed. The fact that diphtheria could be controlled but had not been was viewed as a completely unacceptable situation. Wynne believed that more extensive and intensive methods were needed. As a result, health department officials began to speak of eradicating diphtheria rather than simply controlling it:

> With the ammunition now at hand to completely control diphtheria it is not sufficient to be able to report that progress has been made. The time has come, it is felt by all public health officials, to eradicate diphtheria. The records of the Health Department for the ten years since the introduction of toxin-antitoxin injections reveal that such a result can be accomplished through awakened interest and intensive effort. . . . The fact that diphtheria is still the foremost fatal disease in children from three to five years of age, ranks third in fatality among the ailments of children from one to three, is a basis for urging increased employment of toxin-antitoxin immunization.

The health department then went on to make its goal quite plain: "The aim of the Health Department is to wipe out diphtheria. With the help of the public, the private physician and all other health agencies, this can be accomplished."[22]

The health department could not eradicate diphtheria in New York City on its own. In 1927 more than eighty-two thousand children had been immunized by the department with 191,310 injections of toxin-antitoxin. This work was done in the public and parochial schools, in baby health stations, and in the borough offices of the health department. Though the numbers were impressive on the face of it, with only six doctors and six nurses working full time on immunizations, it would not be possible for the health department to reach all the children in the city, especially those of preschool age. As other cities began to achieve reductions in the morbidity rate due to diphtheria by employing the methods developed by Park, New York City officials were increasingly chagrined at the poor results of the diphtheria-prevention program in the city.

In the fall of 1928 the editors of the *New York Medical Week* joined with the health department in condemning private physicians for their failure to employ immunizations with all their patients. The editors chastised physicians for failing to give the program their full support:

> *The indifference which many private practitioners manifest toward prophylactic measures of this kind cannot be too strongly deplored. The public rightly desires to take advantage of every known method of disease prevention. The state, with equal propriety, seeks to invoke every proven means of safeguarding the community health. The family physician is the logical agent to perform the various preventive inoculations that play so important a part in modern medicine. If he shirks this obligation, through indifference or expediency, he cannot complain if the governmental health forces assume it to his eventual loss.*[23]

The editorial spoke to the implicit threat of the health department appeals to physicians up to this point: if private medicine failed to take up diphtheria prevention, then public medicine would. The position of the health department was made more explicit in late winter 1928. Commissioner Wynne addressed the issue of physician support when he laid out his plans for a massive drive to immunize all young children in the city. In cooperation with the county medical societies, he insisted that every effort be made to have the work done by private physicians. The editors of the *New York Medical Week* were quick to note that the representatives of the county medical societies had been enlisted to participate in the planning of the campaign and had pledged the support of all their members. The editors further stated that in the past, physicians had frequently criticized the health department for "what they have considered an invasion of their proper function." Since physicians had participated in the planning of the new campaign, the editorial continued, the department was giving these critics "an opportunity to show that they are prepared to accept their responsibilities and help in the immunization of all children between nine months and ten years of age."[24]

In one sense the editorial was correct. Wynne was a supporter of the idea that immunizations were the kind of preventive health work that most properly should be conducted by private physicians. Yet if they failed to act, he was prepared to have the department of health step in and do whatever was necessary to prevent diphtheria in the city. He was also prepared to make it known publicly that it was physicians, not the city's health officials, who had failed to protect children against diphtheria.

"Banishing the Scourge of Childhood": The Diphtheria Prevention Commission

On January 12, 1929, Thomas W. Lamont, financier at the J. P. Morgan Company and treasurer of the New York Tuberculosis and Health Association, chaired a meeting of prominent businessmen at the New York Harvard Club to declare "a vigorous war on diphtheria." Lamont wanted to discuss a new campaign to reduce the "alarming rate of increase of diphtheria in New York City."[25] This meeting was called to launch the newly formed Diphtheria Prevention Commission, which Lamont headed. The commission was composed of fifty prominent philanthropists, educators, businessmen, and social welfare activists.[26] The commission also established a technical consultation board, composed of nine of the city's leading pediatricians and representatives of the five county medical societies.[27] Thirty-two of the city's newspaper editors formed an editorial consultation board as well.

Commissioner of Health Wynne spoke at the opening event, where he announced that the commission had enlisted the support of more than eleven thousand physicians in the city, including the five county medical societies, the New York Academy of Medicine, the New York State Charities Aid Association, the Metropolitan Life Insurance Company, the New York Tuberculosis and Health Association, the Milbank Memorial Fund, and the Welfare Council of New York City. The goal of the campaign, Wynne asserted, was to "put an end to [diphtheria], long dreaded for its high mortality rate, within two years in New York City."[28]

Wynne reported that there had been over 13,500 cases of diphtheria in the city during the previous year with more than 700 deaths. The death rate of 12 per 100,000 population was much higher than any city in the state. At that time the health department was making a minimum of four visits to each reported case of diphtheria in the city, for a total of 54,000 visits per year. If every child were immunized, Wynne claimed, the savings to the city could be used to fight other diseases.

Wynne outlined the details of the campaign. It would begin by making toxin-antitoxin injections available free for all those who could not afford to pay for them. Temporary immunization stations were to be established across the city in social service agencies, and all existing public and private facilities for the immunization of children were to be opened up to the general public. Educating the public about the need for diphtheria immunizations was to be a central feature of the campaign. Seven million pieces of literature were to be distributed across the city, especially to the parents of the city's 1.5 million children. In addition, the city's newspapers had agreed to carry on a publicity campaign in support of the commission's "war on diphtheria."

The Diphtheria Prevention Commission's work was supported by grants from the Metropolitan Life Insurance Company and the Milbank Memorial Fund.[29] The Milbank Fund saw the commission's work as an extension of the successful work conducted earlier in the Bellevue-Yorkville Demonstration project. It is apparent that the commission's plan to increase the accessibility of immunizations and its education program were drawn from the reports of the demonstration project.

Wynne concluded the opening meeting by assuring those present that the immunizations were "simple, painless, safe and lasting, and require only a few minutes each time of three visits." A few days later the New York Times gave its full support to Wynne's campaign and to the establishment of the Diphtheria Prevention Commission. The Times editorial emphasized that health officials had to depend on the public's cooperation in the control of diphtheria since it "has no such autocratic powers in the prevention of diphtheria as it has in the prevention of smallpox." (The editorial was referring to the fact that smallpox vaccination was mandatory under the city's Sanitary Code.) Without such police power, the editorial noted, it was necessary for educational and charitable agencies, the life insurance companies, and the press together to bring home to the people, and "particularly to mothers the danger that lies in the spread of the disease and the ease with which it can be avoided."[30]

Wynne's campaign for diphtheria prevention, like Hermann Biggs's campaign for antitoxin, would use persuasion, rather than coercion to achieve its goals. As with Biggs's justification for the bacteriological diagnosis program, Wynne also promised that immunization would save the city money. While both campaigns stressed the need for cooperation among many different constituencies in the city, the focus on mothers in the immunization campaign was a unique feature.

With support from all quarters, the Diphtheria Prevention Commission's work proceeded at a dizzying pace. Within weeks after the commission was established, forty-four new immunization stations were opened across the city. Dr. Edward Fisher Brown, the executive director of the commission, reported that by the end of January forty to fifty letters per day were pouring in from parents inquiring about the antidiphtheria treatments.[31] By the end of the month the "war" on diphtheria was being waged with full force.

Susan Sontag and other critics and historians have argued that the invocation of "war" and other military metaphors in public health campaigns has a long history. In the wake of the germ theory, the use of such metaphors achieved new credibility and precision.[32] These metaphors invited characterizations of diseases as an invasion of a society and efforts to reduce mortality as a fight, struggle, or war. Early in the campaign against diphtheria, similar characterizations appeared. A cartoon printed on the editorial page of the New York Evening World, and reprinted in the health department's Weekly Bulletin, depicted diphtheria as a large skeletal specter draped in a dark cloak, looming over a white man named "Father Knickerbocker," who is clothed in colonial garb and carries a large spiked club. Father Knickerbocker, as the representation of the early settlers of New York, glowers back at the specter while protecting two small children who cower in fear behind him. The caption reads: "Your Fight Too, Fathers and Mothers," implying that public health officials, as representatives of the city, cannot protect children alone. Like the popular melodramas of the day, diphtheria had become the enemy that only the power of medicine could defeat.

War-making metaphors also contribute to the stigmatization of illness and disease. Since diphtheria was endemic in New York City, little stigma had been attached to outbreaks or deaths associated with it. Yet now the rhetoric of the campaign cast blame upon parents and physicians for the disease. The very presence of diphtheria became a synonym for neglect. In public speeches at the start of the campaign, Wynne expressed this in harshly moral terms: "Diphtheria is entirely due to the indifference and innocent ignorance on the part of parents."[33] Physicians too were at fault in Wynne's view:

> That diphtheria has persisted in New York is due to the fact that the medical profession has failed to take aggressive steps to prevent this disease in the families under their charge. Doctors may well hold themselves responsible for every case of diphtheria among their patients if they have not exercised every means to protect their children with toxin-antitoxin.[34]

The use of military metaphors in the control of disease also constructs enemies, and in this case the two primary ones were the disease itself and uncooperative physicians. Parents were a secondary enemy. Wynne's placing of blame on physicians was a thinly veiled reference to the longstanding political and professional disputes between physicians and the health department over the authority to define appropriate interventions in the prevention and treatment of diphtheria.

Physicians and the Diphtheria Prevention Campaign

While parents had little ability to respond directly to the charges made by Wynne, physicians did respond. At the same conference where Wynne pointed to the failures of physicians, Dr. Nathan Van Etten made an apologetic speech from the perspective of the private physician. He noted that since only 5 percent of the average practice was concerned with communicable diseases, this fact should induce "tolerance and patience with his slowness to rise to the highest co-operation in a campaign for the prevention of diphtheria." Though physicians had known about toxin-antitoxin for fifteen years, Van Etten continued, they needed to feel their way before employing it by studying the statistics on immunity as revealed by scientific research. Even so, he remarked, not all physicians had been slow to respond to the diphtheria campaign. Pediatricians and general practitioners were ardent supporters, whereas those in other specialties not related to children had been more reluctant to get involved. Finally, Van Etten deflected attention away from the issue of physician neglect by echoing Wynne's reproach of parents. He declared that ultimately parents alone would be responsible for protecting their children: "The educated parent will be held directly responsible when a child dies from diphtheria if he shall have failed to ask for protection by toxin-antitoxin."[35] Van Etten's poor apology expressed what physicians had long argued, that they had to make their own medical assessment of the value of the intervention the health department proposed. His position that physicians' resistance to immunizations was due to the lack of parental demand for it seemed unduly gratuitous, though other physicians would make such claims as well.

Dr. Charles Herrman also blamed parents' obstructionism. Commenting on the health department's immunization program, he claimed that only 10 percent of the mothers in his practice recognized its value and "were glad to have their babies immunized." An additional 10 percent consented after a little explanation and persuasion. Overall, he found that a large number of his patients preferred to wait. "Some told me frankly that they did not believe in vaccination," he wrote, "but they had it done because the child could not be admitted to school without a vaccination certificate." Hermann attributed his patients' resistance to immunization to "cults" such as antivaccinationists, antivivisectionists, and antiimmunizationists.[36]

Despite such rationalizations, Wynne continued to focus on the unresponsiveness of physicians in his public remarks. While his attitude could be interpreted simply as a reflection of his desire for all parties to take up the campaign with the same zeal he expressed, it was more likely an expression of the ingrained institutional mistrust, if not hostility, between the health department and private physicians. In March 1929 two children reportedly died of diphtheria in the city without receiving proper medical care. Wynne eagerly seized the opportunity to announce that health department physicians would be sent anywhere in the city at any time, day or night, to attend anyone suffering from diphtheria.[37] This represented a major departure in the department's policy established under Biggs of not interfering with patients of private physicians. In addition, Wynne announced that doctors in the health department's clinics would be allowed to go to private homes and immunize children when parents notified the department that they were unable to bring their children to the clinics. These were bold moves on Wynne's part, though not unexpected. The health department's diphtheria program had come into existence and been maintained in part because its leaders were willing to supersede functions traditionally associated with private medicine, and as a result, a "natural" antagonism had grown up between the two.

Physicians' responses to the specific practices associated with the campaign are best viewed in the context of their attitude toward public health and prevention. In the early decades of the twentieth century, physicians in urban areas found their economic position deteriorating despite increasing professional respectability. They expressed increasing bitterness over the number of free medical services provided to low-income groups by local and state boards of health or by private social welfare agencies. Physicians, preoccupied with their own economic interests, failed to see that only a relatively small percentage of the working poor could afford private care. They failed to appreciate the health needs of the vast majority of this large segment of the population who were the legitimate concern of the health department and private agencies.[38]

An even greater encroachment into the private physicians' domain occurred in May, when the Diphtheria Prevention Commission sent circulars to forty-five hundred physicians in the city asking them to agree to a standard maximum charge for diphtheria immunizations of two dollars for each injection.[39] The letter, co-signed by the president of the New York County Medical Society, was ostensibly drafted after the health department received reports that many parents refused to have their children immunized because they feared the high fees that they would be charged. The radical experiment of setting a fixed price for immunizations garnered support from more than 41 percent of the city's physicians, who agreed to set aside one day a week for diphtheria treatments.[40] The free immunizations provided by the health department were intended for parents who could not pay, but many parents refused to avail themselves of this service because it was perceived as charity, and they preferred to pay a private physician.[41] The health department wanted to avoid any perception that services for the poor were being used by those who could afford to pay for immunizations.

Shortly after the diphtheria campaign opened, an editorial in the *New York Medical Week* commended the health department for the efforts it had taken to encourage physicians to immunize their patients and to push parents to ask for such immunizations from private physicians. They viewed this as an important conciliatory step on the part of the health department, aimed at mollifying the past resentment:

> *Heretofore doctors have resented their exclusion, through undiscriminating municipal competition, from a form of practice which they are competent to assume. . . . The present plan not only enables the private physician to participate in disease prevention, but employs the prestige of official and semi-official bodies to make him the nucleus around which this type of work shall revolve.*[42]

The editors saw the campaign as an opportunity for physicians to demonstrate their concern about preventive medicine. The editors frankly acknowledged, "If it [private medicine] fails to cooperate, it cannot complain if the state assumes the obligation it has ignored. The work must be done."

Wynne's tactics were perceived by physicians as a move toward the expansion of the role of the state in medical care. Physicians' resistance to these tactics is evidenced by the reassurance that the editors of the medical journals gave to their readers in their commentary on the campaign. The editor of the *New England Journal of Medicine,* for example, wrote that the fact that New York's diphtheria campaign explicitly recommended that immunizations be given by private physicians proved that it was not a harbinger of "State Medicine."[43] The editors of this journal, along with other physician leaders, noted that physicians needed to pay more attention to preventive medicine. It appears that Wynne's inclusive actions towards physicians had some small impact on mediating his more aggressive tactics.

Educating the Public

The Diphtheria Prevention Commission presented its project as educational, aimed at alerting the public to the dangers of diphtheria and the value of immunization. Education, as Progressive Era reformers promoted it, was primarily a means of shaping individual behavior. In this case the goal was to convince parents, especially mothers, that immunization of their infants and preschool-age children was safe and necessary. The reduction of the "fears" of mothers became one of the most visible aspects of the campaign. Dr. Lee K. Frankel, senior vice president of the Metropolitan Life Insurance Company, argued that "before the goal of perfect freedom from the disease could be achieved, mothers' clubs, parent and teacher organizations, and schoolchildren must be enlisted in a campaign against 'fear.' " "It is fear," continued Frankel, "that is in the minds of mothers when they refuse to allow their children to be immunized, and before we can overcome that fear with knowledge we must get into the homes. Fundamentally the campaign is an educational one."[44]

Beginning in the early decades of the twentieth century, executives at the Metropolitan Life Insurance Company (hereafter Met Life) moved to support public health projects in New York City as part of its attempts to improve the image of the insurance industry in the city. The company understood the notion of going into the home to educate parents about the value of immunization quite literally. In 1909 Haley Fiske, president of Met Life, hired Frankel, who was at that time director of the United Hebrew Charities, to head Met Life's new Welfare Division. This division was established to provide direct medical care to Met Life policyholders, especially immigrants living in the crowded neighborhoods in New York City. Frankel stated that with the new division, Met Life had moved into "a new era which joined in the battle against disease and in which it hoped to lead the world in preserving lives."[45] Though espousing humanitarian ideals, the company's new work was not entirely altruistic. It hoped to bring about both lower mortality rates and increased insurance sales among the working-class population.

Frankel was an early and vocal supporter of Dr. Wynne's diphtheria-prevention campaigns. Beginning in 1923, his division had prepared numerous pamphlets about immunization that advised parents to have their children inoculated. These pamphlets employed simple picture stories to instruct parents and children about diphtheria immunizations. They were distributed free to Met Life customers. But the company's role in the diphtheria campaign went far beyond simply providing educational materials. It also cooperated with the health department in establishing and organizing immunization clinics across the city. Agents carried consent forms for immunizations to the homes of prospective clients, and in some instances they even made the contact between the children and the clinics.

Child-saving work enhanced the company's image and helped it reach potential customers. Its statisticians reported a decrease in the diphtheria death rate among its policyholders from 15.5 per 100,000 in 1923 to 12.8 per 100,000 in 1924. Frankel attributed this drop to the increasing numbers of children that had been immunized through the company's efforts.[46] During the diphtheria campaign in 1929, Met Life placed all of its existing materials and clinics at the disposal of the health department.

Media Blitz

Media campaigns against disease always serve to establish a framework of expectations that shape the way individuals, communities, and institutions respond to disease.[47] By the summer of 1929, no one living in New York City could fail to be "diphtheria conscious." The city was inundated with material about the disease. Information was splashed across every public conveyance. For Wynne, a high level of publicity was absolutely necessary to the success of the campaign: "This idea of diphtheria immunization had to be 'sold' almost in the same manner as chewing gum, a second family car, or cigarettes. This meant reaching a population of six million people."[48] And sold it was. One hundred thousand placards were placed in all the transit vehicles in the city: in subways, surface cars, and elevated trains. These placards implored "Save Children from Diphtheria!" in bold letters, emphasizing that the disease was a "dangerous deadly disease of childhood" but that it could be wiped out by "safe, harmless and lasting" immunizations.

There were significant differences between the newspaper coverage that supported the introduction of diphtheria antitoxin in 1894 and the media campaign conducted by the Diphtheria Prevention Commission in 1929. In

1894, only newspapers were involved in publicizing antitoxin; whereas in 1929, information about diphtheria was disseminated through every form of mass communication available—newspapers, radio, billboards, flyers, and religious bulletins. In 1929, the media campaign could more rightly be called an advertising campaign for the commission's activities.

The 1929 campaign circumvented the objective reporting function of the press by coopting editors of the major newspapers into service on the commission. The reporting of the campaign was thus more public relations than news.[49] Newspapers did not even make a pretense of criticizing the merits of the campaign or the risks involved in such a massive program of active immunization. In 1894 the health department's link to the newspapers had been a more covert one. It provided technical information to the papers and, as discussed above, published scientific reports about the merits of antitoxin. At the time, such stories served to sway a suspicious medical community that had fewer sources of information about advancements in medicine. There are, however, some similarities between the campaigns as well. In both periods the rallying cry that science had robbed diphtheria of its terrors raised expectations and served to obscure any critical review of what science promised.

In the absence of an epidemic, the media campaign was used to create a heightened sense of urgency about the need to have young children immunized against diphtheria. The commission took advantage of this atmosphere by providing unprecedented access to facilities where the immunizations could be obtained. By May 1929, the city opened thirty immunization stations in the public parks as well as an additional forty-four new clinics. These stations were kept opened throughout the summer in the city's park buildings or in tents adjacent to them. Dr. Béla Schick, inventor of the Schick test, who had been appointed to the technical consultation board established by the Diphtheria Prevention Commission, noted that only the first of the three injections needed to establish permanent immunity against diphtheria would be given to children in immunization stations in the parks. The children were then given a card directing them either to the health department or to a private physician for the two remaining injections.[50] Thus in creating these new sites for immunization, the commission tried not to undermine the position of the private physician.

In July the city deployed six renovated snow-removal trucks as "health-mobiles." These mobile immunization stations were equipped with specially built refrigerators to store the toxin-antitoxin and staffed by two nurses and one physician to administer the injections. The healthmobiles toured the "congested" sections of the city, stopping to allow the nurses, who were fluent in at least two languages, to deliver talks about the dangers of diphtheria. The trucks were outfitted with circus paraphernalia and balloons to attract the children. Edward M. Brown, the commission's director, described the new strategy: "We are taking a page from the history of the old Indian medicine doctors and putting it to use in the field of modern public health work. . . . This will be the first time in the history of public health work that clinics on wheels are to be used."[51] Along with the healthmobiles, clinics were set up on the city's beaches as well as on the St. John's Guild Floating Hospital, which had served the city's poor children since 1874.

New Yorkers simply could not fail to see or hear information about diphtheria in 1929. The local gas and electric companies included circulars in their monthly billing. Leading department stores in the city placed signs in their store windows and in their newspaper advertisements. Articles on the campaign appeared in all of the city's thirty-six daily newspapers, including those published in ethnic communities. In order to reach all of the ethnic groups, the health department had its literature translated into ten foreign languages and established a special Foreign Groups Committee composed of prominent members of the city's ethnic communities to advise the commission.[52]

Endorsements were also sought from civic and religious leaders in the city. Cardinal Hayes sent out a special pastoral letter, which was read at masses at all the Catholic churches in the city; and four hundred thousand copies of it were distributed to parishioners. The text read in part:

> A matter that so deeply affects human life, family happiness, and the health of the community, is of necessity of a great concern to your Archbishop. . . . [W]e have decided to urge through our pastors the wisdom of adopting scientific means of preventing diphtheria, and to arrange through our parochial schools to put this means at the disposal of those unable to patronize their family physician.[53]

Given the city's large Catholic population, Cardinal Hayes's support was significant. The study conducted in the Bellevue-Yorkville project indicated that the support of priests and clergy was very important in persuading some families to allow their children to be immunized. The commission also attempted to address the concerns of specific ethnic groups by using prominent members of those communities as spokespersons for the campaign. The Foreign Groups Committee canvassed in specific ethnic neighborhoods. The Comite Puertorriqueno, for example, arranged special meet-ings, conducted house-to-house canvasses, and distributed leaflets at fra-ternal organizations and neighborhood festivals in Puerto Rican neighbor-hoods.[54]

The health department's own tabloids featured comic strips with mothers and babies. Some featured stories of mothers from the preantitoxin era, shown crying over their sick children, who were dying from diphtheria. Short commentaries directed to mothers, written by Wynne and other prom-inent members of the Diphtheria Prevention Commission, appeared regu-larly. The text in these tabloids cast diphtheria as the "enemy" of children that was "stalking" their lives. Parents were urged to be responsible protectors of them joining with the "war" being waged against the disease by physicians. Such language was also used in the regular radio broadcasts about the cam-paign. The Tower Health Talks, sponsored by Metropolitan Life and broad-cast from its main office in the Metropolitan Tower in Manhattan, featured health department officials such as Park and Wynne.

An official story of the history of diphtheria emerged in the pages of the tabloids and in these talks. A good example of this story is a radio talk given by Dr. Iago Galdston during the campaign. Galdston described the tragedy of a family living in Manhattan during the colonial period who had lost a child to diphtheria. Not even the great Samuel Bard, New York's famous co-lonial physician and author of a treatise on diphtheria, had been able to save the child. Only the advances of modern science could prevent children from dying from diphtheria, Galdston recounted. He described in simple language the discovery of the Loeffler bacillus and diphtheria antitoxin. "But since modern medicine is not satisfied with merely finding the cure for a disease," Galdston stated, with the development of toxin-antitoxin, "it wants to go further. It aims at prevention rather than cure . . . By the univer-sal application of this preventive, diphtheria can be wiped out completely and can be made as rare as the plague or leprosy." He then went on to explain how toxin-antitoxin produced immunity. Galdston concluded with the by-now-familiar comment to his listeners: "The means for preventing this are all at *your* command and you should use them for your sake and the sake of *your* children" (emphasis in original).[55]

Galdston's story, like much of the literature produced by the Diphtheria Prevention Commission, emphasized that the disease had been present in New York City from its earliest days. The familiar heroic physician figure, in this case Samuel Bard (but often William Park or Béla Schick), was cast as the protector of children against the disease. Medicine was cast as having failed in the past, but now, in the age of modern science, as having conquered the disease. The moral of the story was that the final defeat of diphtheria depended on parents.[56] Galdston's story and ones similar to it firmly established the history of diphtheria as the triumph of science. Science had conquered fear, ignorance, medical failures, and, of course, diphtheria itself. The final exhortation to parents to trust in the ability of science left little room for doubt, skepticism, or resistance.

These radio talks and tabloids were designed both to frighten and to motivate parents to have their children immunized. Though the intent was to reduce the parents' fears of immunization, the use of language and imagery that foregrounded the terrors of the disease in some ways had the opposite effect. Commissioner Wynne had to admit that much of the literature exacerbated the fears of mothers and of the public rather than reduced them. "A nomenclature with special reference to the appeal of mothers must be devised," Wynne noted during the first few months of the campaign, "less terrifying, more subtle, which specifically answers the definite prejudices mothers still have against immunization."[57] Wynne also noted that the campaign had revealed widespread prejudice against immunization among fathers, many of whom had unfortunate experiences with inoculations in camps during the war. Though Wynne made note of these concerns, it is not clear how they were addressed in any concrete way. There was no discernible change in the literature produced by the Diphtheria Prevention Commission over the course of the campaign that addressed this question.

The public relations role played by the press in diphtheria prevention masked important medical questions about active immunization. No newspaper questioned the use of toxin-antitoxin as the immunizing agent. Few among the public would have known that an alternative agent, toxoid, existed. As the editors of the British Medical Journal noted, "[In] the large amount of forceful propaganda in the present campaign in New York, aimed at the eradication of diphtheria within the next decade, the use of toxin-antitoxin mixtures is urged and toxoid is scarcely mentioned."[58] Physicians and public health experts were aware, however, of evidence suggesting that toxoid was a better immunizing agent than toxin-antitoxin and produced fewer reactions in older children and adults.[59] Park was not unfamiliar with this issue. The health department did shift to using toxoid for younger children by 1931, but Park felt that toxin-antitoxin was best for older children despite the reactions, because the immunity produced with it lasted longer.[60]

The literature produced by the commission did not note these reactions or the possibility of accidents from the improper use of toxin-antitoxin mixtures.[61]

Another issue received little notice: many children in the city were being given toxin-antitoxin without an initial Schick test. Park argued that after the experimental phase of the immunization work, the need for a primary Schick test was reduced. "The decision simply depends on whether in any given age group and locality it is more trouble and annoyance to make a Schick test and so omit unnecessary inoculations, or to omit the Schick test and inoculate all children," he wrote.[62] Park based his decision on the data he had generated from the thousands of schoolchildren that had been immunized in the city.

Again, there were dissenting views. One British public health officer wrote that the abandonment of the primary Schick test was "unsound and unscientific," as it was based only on empirical evidence. He continued, "What right do we have to inject a patient who does not need it with any drug or bacterial product?"[63] Park's reputation, the successful immunization trials, the absence of any accidental deaths in New York City due to toxin-antitoxin, and the failure of the press all combined to downplay questions about active immunization during the campaign.

By the end of 1929 a total of 292,000 children were immunized with toxin-antitoxin, and the diphtheria death rate fell from 6.75 per 100,000 in 1928 to 2.75 per 100,000 by the end of the year. The bulk of the immunizations were given by the health department, and only 30 percent by private physicians.

The impact of the campaign on the incidence of diphtheria in the city was not immediately apparent, since it took up to six months for the immunizations to have an effect. Comparing the number of cases reported in the third quarter of 1929 with those in the same quarter in 1928, it was found that there were about 200 fewer cases. Overall, the number of cases had steadily decreased from July to December during the year as the number of immunizations increased. However, given that the prevalence of diphtheria followed a cyclical pattern of recurrence, it was expected that 1929 would be a year with a comparatively low number of cases. Edgar Sydenstricker, director of the Division of Research of the Milbank Memorial Fund, predicted that diphtheria in New York City peaked about every six-and-a-half years. The last peak had occurred in 1927; therefore, 1929 was not expected to be a year with a large number of cases. Comparing the number of cases in 1930 to the last low year, 1926, Sydenstricker found that in 1930 there *was* a substantial reduction in both cases and deaths, which he attributed to the immunization campaign.[64]

The campaign was declared a success even though the goal to "eradicate" diphtheria was not reached. The number of cases were cut by 56 percent and deaths by 80 percent when it ended in 1931.[65] A total of 522,243 children were immunized between 1929 and 1931. The health department estimated that seventeen thousand children escaped diphtheria during the course of the campaign, had the average case rate for the last ten years prevailed during the time of the campaign.[66] In 1930 the Diphtheria Prevention Commission was the recipient of the Harmon Award, which was given for the "complete record of a well planned and executed program covering a year's work in publicity carried on by a public or private agency engaged in social or health work submitted by an agency in a city or country of 200,000 or more population."[67] The work of the commission was taken over by the health department in 1931, and immunizations against diphtheria continued as a routine part of its activities.

Conclusion

The efforts to control diphtheria in New York City did not end with the passing of the Diphtheria Prevention Commission in 1931. While its notable successes included making the city's population "diphtheria conscious," it did not succeed in making private physicians take on providing immunizations at reduced rates on a regular basis. A substantial number of children were immunized, but with births averaging 125,000 per year in the city, only a regular, institutionalized program of immunizations could keep the incidence of diphtheria at a low level or at best eradicate it. By 1933, the editors of the New York Times reported that more than 700,000 children in the city had not been immunized.[68]

The stated goals of the Diphtheria Prevention Commission were to create a "diphtheria conscious" public, to immunize as many children as possible, and to reduce the mortality and morbidity from the disease. The directors of the campaign measured its success with regard to those variables. Indeed, more children were immunized than at any previous time, the public was certainly made conscious of diphtheria, and mortality and morbidity were reduced. The strategies used in the campaign had inadvertently revealed underlying keys to achieving such success in public health campaigns. First, economic barriers were reduced by making immunizations affordable and accessible throughout the city in an unprecedented way. Second, the campaign cut across class and ethnic lines, making the immunization of all children a visible community concern. Third, cultural and language barriers were addressed by producing literature in the languages spoken by all the city's ethnic groups and by the use of health care professionals from ethnic groups in those communities.

Physicians, however, did not continue to immunize children at the same level once the commission was dissolved. When healthmobiles were removed, clinic hours were shortened and alternative immunization sites were closed. Funds for translation of educational materials were reduced, and in the absence of an epidemic outbreak of the disease, the impetus for continued public support of immunization waned and diphtheria once again receded into the background.

Historians of the Progressive Era generally tout public health campaigns of this period as the most successful among the variety of reform movements of the time. Michael Katz argues that part of this success was due to a public health campaign's ability to lower mortality "without improvements in macroeconomic conditions, such as unemployment, or structural changes in the distribution of wealth and power." In addition, he argues that since public health was not class specific, it "sidestepped the contradictions between public responsibility and privatism by extending the reach of the state without singling out any group as inferior or degraded."[69]

The diphtheria-prevention campaign shows the profoundly class-bound nature of one public health reform effort. The poor and working poor were immunized at different sites from those who could afford private care. Nor were the sites for the poor intended to be permanent ones. While nearly half of the city's physicians provided immunizations at reduced cost during the campaign, there was no attempt to establish reduced fees on a permanent basis, which physicians certainly would have resisted. The rhetoric of the campaign, emphasizing maternal irresponsibility, was directed at poor immigrant women, although it certainly had an impact on all mothers in the city. Their fears were treated as irrational and unfounded rather than valid expressions of maternal concern. The so-called fearful and resistant mother was used both materially and rhetorically as a scapegoat in the ongoing conflict between physicians and the health department in much the same way as the so-called welfare mother functions in current debates in health policy.

The fact that no group or class was singled out as inferior or degraded in the antidiphtheria campaign has more to do with the behavior of diphtheria than with any desire on the part of public health leaders to avoid stigmatizing particular groups of people. Diphtheria was just one of the contagious diseases of children that had long been present throughout New York City, but especially in the more crowded neighborhoods. It did not suddenly appear in these neighborhoods, causing people to fear its victims, and there were no unusually large outbreaks of the disease during this period. There were no separable, distinct populations of those immune and susceptible to the disease. Everyone was exposed to infection in streets, stores, schools, and public vehicles.

This final public effort in diphtheria control in New York City illuminates the continuing tensions between public health and private medicine that each aspect of the diphtheria-control program in New York City had brought to the fore. Historians have noted that in this period of the "New Public Health," health departments had retreated from efforts that focused on the environment and on practices of isolation and disinfection to an emphasis on health education and personal hygiene, specifically on the "use of the physician as a real force in prevention."[70]

The diphtheria prevention campaign shows that public health leaders' desire to use physicians as a force in preventive medicine was an ambivalent desire at best. This ambivalence was reflected in the use of persuasion undergirded by the threat of coercion in the efforts to encourage physicians to participate in the immunization campaign. While opportunities were provided by the health department to address its economic concerns by subsidizing immunizations and promoting the use of private physicians, public health leaders also made it clear that they would exercise their authority in disease control if private physicians did not immunize children when called upon to do so. Physicians, on the other hand, continued to focus on the potential of the health department–controlled interventions to violate the boundaries between preventive and curative medicine rather than on the interventions themselves.

While there was wide recognition on both sides of the fragility of the boundary between preventive and curative medicine, this final public diphtheria campaign shows that public health leaders' hope that a shared commitment to the scientific value of the interventions proposed would obviate tensions between the two groups was not realized. From its inception, the architects of the diphtheria-control program used the fact that while the interventions they proposed obviously threatened physician autonomy, this threat was less important than the value of the interventions, which were the products of a supposedly value-free laboratory science. The laboratory had made diphtheria impersonal, but it could not make the means to control it apolitical. Large public disease-prevention campaigns have both positive and negative effects. By its success, the diphtheria-prevention campaign produced and strengthened a societal commitment to immunization by fostering the notion that it was the civic duty of parents to have their children immunized. On the other hand, it diverted resources, emphasis, and attention away from the public and private costs and the institutional and social structures needed to maintain the immunization levels of all children on a routine basis.[71]

The ideology of the "New Public Health" as expressed in New York City was sustained by a commitment to dispense the benefits of science to the widest possible public. The rhetoric of the campaign heavily promoted immunization as the right intervention because it was a scientific solution to diphtheria control. The diphtheria-prevention campaign is evidence that physicians did not interpret a commitment to "science" in the same terms as did those involved in public health. Thus, in the end, science did not triumph in any simple way over the diverging interests of public and private medicine in the control of diphtheria.

In the next decades after the 1930s, the specific problems associated with sustaining and maintaining appropriate levels of immunization against diphtheria faded into the more general problem of maintaining immunizations for an increasing list of childhood infectious diseases that the work on diphtheria had made possible. From this period forward to the present, public debate about childhood immunizations would center on the issues that the diphtheria-prevention campaign brought to the fore: educating parents about the benefits of immunizations, the role of private physicians in providing them, costs of the vaccines, access and availability of sites for the provision of immunizations, and the role of local and national government in support of each of these factors.

EPILOGUE

Disease is largely a removable evil. It continues to afflict humanity, not only because of incomplete knowledge of its causes and lack of individual and public hygiene, but also because it is extensively fostered by harsh economic and industrial conditions and by wretched housing in congested communities. . . . The reduction of the death rate is the principal statistical expression and index of human social progress. It means the saving and lengthening of the lives of thousands of citizens, the extension of the vigorous working period well into old age, and the prevention of inefficiency, misery, and suffering. These advances can be made by organized social effort. Public health is purchasable.

HERMANN BIGGS, 1911

The successful immunization of hundreds of thousands of preschool and school-age children in 1930 led many observers to believe that diphtheria would soon be a thing of the past. Health officials envisioned a rosy future in which humanity would be free not only of diphtheria but of all infectious diseases. In many respects they were right. Improvements in immunizations for diphtheria and new immunizations for whooping cough, measles, and mumps led to impressive decreases in the incidence of these diseases throughout the population. Between 1980 and 1995, only forty-one cases of diphtheria were reported in the United States.[1] Drugs and vaccines, along with improved sanitation and nutrition, had seemingly defeated the dangerous microbes that were an ever-present reality at the beginning of the century. Chronic diseases such as cancer, cardiovascular disease, and mental illness moved to the center of medical and popular attention. Complacency about infectious disease settled in. As the infrastructure for monitoring and treating these diseases was dismantled, physicians viewed the study of infectious diseases "as a dead-end as a vocation and an increasing yawn as an intellectual discipline."[2]

Yet by the late 1980s the future seemed far less bright than Hermann Biggs, William Park, and Shirley Wynne had envisioned. The advent of the pandemic of human immunodeficiency virus (HIV), which is associated with the production of acquired immune deficiency disease syndrome (AIDS), shattered our complacency. AIDS had at last count reached over 5.8 million people worldwide, with over half a million infected in the United States alone.[3]

The AIDS epidemic refocused attention on the lessons learned in earlier confrontations with epidemics of infectious diseases. Scientists, journalists, the public, and historians all searched for analogies with past diseases to answer pressing questions of the day. Why had AIDS emerged when and where it did? How did the disease spread among members of particular groups? Most importantly, what does the history of medical science and public health in this century suggest about our ability to control the epidemic and eventually to cure the disease?[4] Although such questions have spurred renewed interest in the history of medicine and public health, they also put historians in the difficult position of using the past to explain the present. Historians are mindful that "history is not a predictive science." What the past can tell us about the present is always contingent and partial.[5]

With respect to this study, I must first note the obvious: diphtheria is not AIDS. Diphtheria evoked societal anxieties about protecting children, about the appropriate application of scientific and medical knowledge, and about the proper boundaries between science, public health, and the state. Along with these issues, AIDS has also triggered anxieties about sexuality, sexual orientation, drug abuse, and other so-called deviant behaviors. Diphtheria raised none of these issues. Diphtheria was a disease that ravaged innocent children, whereas AIDS strikes the innocent as well as those deemed guilty for engaging in dangerous behaviors.

Furthermore, the historical moment when diphtheria was brought under control cannot be recaptured, for institutional and societal structures and American society itself have all fundamentally changed since that time. It is no longer possible for a municipal health department to singly play a pivotal role in the control of a disease as the New York City Health Department did with diphtheria at the beginning of this century. In particular, developments in the institutional shape of the medical profession, the growth of the pharmaceutical industry, the evolution of the federal drug regulatory infrastructure, and the changing role of public health agencies all have made disease control more complex than could have been imagined at that time.

Yet acknowledging the broad differences between diphtheria and AIDS, it is also true that in different ways both of these diseases reveal our dependence on the laboratory and its findings, and the problematic relationship between medical knowledge and its applications. Diphtheria is situated at the opposite end of the spectrum from AIDS with respect to these issues. The control of diphtheria made evident for the first time the promise and the force of the laboratory in infectious disease control. AIDS, on the other hand, has shown us how deeply dependent we have become on the laboratory and how aware we are of its limitations. In both cases, expectations were high that laboratory sciences could control disease completely. The story of the control of diphtheria, as I have told it, reveals that the application of laboratory and medical knowledge was dependent on many factors—biological, political, and social. The history of the control of diphtheria, as told by those involved, emphasized the *power* of scientific knowledge rather than its limitations, and of technical rather than social solutions to disease control. As we sort out the conflict between the protection of public health and the protection of civil liberties with AIDS, we can more readily see what Biggs, Park, and Wynne could not—that technical solutions for disease control have a variety of complex social implications.

Despite my hesitation to use diphtheria to provide specific lessons for the control of AIDS, in a more general sense there are three aspects of the history of the control of diphtheria that could provide occasion for more serious interrogation of our current AIDS policies: mandatory testing, education campaigns, and government provision of vaccines.[6]

Mandatory Testing

The introduction of mandatory throat cultures for the diagnosis of diphtheria provides an example of mandatory testing of the population. The success of this diagnostic program was due, first, to the distribution of culture kits and of test results to all physicians at accessible locations and at public expense. More important, criticism of the program by physicians and the public was muted when testing was linked to the use of antitoxin, which was also initially provided at public expense. In the case of diphtheria, testing was seen in a more positive light when it became a way to ensure treatment and prevention.

We can now track a similar change in attitude toward AIDS testing, as new more effective treatments have emerged. Initially, calls for mandatory testing for AIDS were widely resisted, in part because there was no effective treatment available, and also because the stigma associated with an AIDS diagnosis resulted in severe social costs. Currently, tensions are easing as new treatments offer compelling reasons to notify and test people who might carry the virus. However, as I will discuss below, testing remains problematic if it is not connected to treatment.

The diphtheria case also shows that accessibility to testing was a factor in its acceptance. We must remember that because the stigma associated with diphtheria was not as severe as that with AIDS, the New York City Health Department did not have to address in any significant way the ethical dilemmas that testing raises today in pitting people's confidentiality rights against the state's duty to inform and protect the public's health.

Education Campaigns

Biggs, Park, and other leaders in the New York City Health Department were well aware that they had to convince a skeptical medical community and the public of a radically new way of understanding disease; therefore, educating the public about diphtheria was an important part of their work. As I have pointed out, the antidiphtheria campaigns did not provide a forum for critical review of proposed interventions to control diphtheria. They did succeed, however, in creating a diphtheria-conscious public. Though it is difficult to directly assess the results of these educational campaigns in a city with a multiethnic population of varying degrees of literacy, what is notable is the concerted attempts to reach all ethnic groups. The recognition that New York City had an ethnically diverse population that required specific strategies targeted to these groups was a singular aspect of the antidiphtheria programs.

Surprisingly, what was recognized as a fairly obvious need for public health programs to address the diverse populations found in New York City in the 1920s seems to have been lost in the decades between 1930 and 1980, when AIDS appeared. Well into the first decade of the AIDS epidemic, educational efforts in communities of color in many urban and rural areas were fragmented and haphazard, as I have discussed elsewhere.[7] Reports indicate that many people in groups that are increasingly at risk—Native Americans, lower-income Hispanic and African Americans—continue to lack some vital facts about the transmission and prevention of AIDS.[8] Treatment and prevention information in accessible language levels and in languages other than English were slow in coming.

Provision of Treatment

I have argued that the public antidiphtheria campaigns were critical to the successful control of diphtheria. It is significant that these campaigns focused on removing the stigma associated with diphtheria and providing access to treatment and preventive vaccines for the entire population at risk. In the case of the campaigns to immunize the preschool population, Wynne and his supporters did not ignore the need to subsidize the costs of these immunizations through public and private efforts.

Again, the contrast with AIDS is striking. Now that viable treatments are available, many are unable to use them because the costs are prohibitive. The federal-state partnership designated to pay for AIDS drugs for the indigent has run out of funds in twenty-five states, so that the epidemic can now be characterized by the "therapeutic-haves" and the "therapeutic-have-nots."[9] This is precisely the kind of social fragmentation and stigma that the public health leaders in the diphtheria story sought to avoid. Biggs's ability to marshal public and private funds for diphtheria antitoxin, for example, was predicated on the argument that the funds would be used to provide relief to the poor. Containing diphtheria among the poor would then directly benefit those more well off. Public financing of diphtheria antitoxin was justified because everyone would benefit.

The ultimate success of the antidiphtheria programs was critically dependent on their being perceived by physicians and the public as classless rather than class-conscious interventions. By contrast, public health and government officials have not yet identified a strategy that can effectively marshal public support and the necessary public funding of drugs for many poor people with AIDS. Although the obvious complexities of our current health care financing crisis cannot be ignored, the lesson from the history of diphtheria is that public financing of drugs for treatment or prevention is key to the successful control of infectious disease that strikes a large part of the population.

Conclusions

In sum, these lessons from the history of diphtheria do not offer simple solutions or easy resolutions to our current problems in the control of diseases like AIDS. These lessons cannot tell us, for example, whether widespread testing of the population is good or bad; nor can the control of diphtheria offer a blueprint for how new treatments should be dispersed to those most in need. This history reveals the deep tensions inherent in the research and policies that we develop to control disease. These tensions between public health and private medical interests, between private pharmaceutical firms and public health, between government and private interests, and between the public's fears of contagion and death and individual rights can only be addressed if we acknowledge them.

Diphtheria was controlled because Hermann Biggs led an effort that combined research in the new science of bacteriology with public health policies that applied that research to the broadest possible population. Though his ability to manipulate the public and the political system with outside money and publicity waxed and waned over the period under study, the search for the perfect control of the disease—the eradication of diphtheria—led his successors to develop a research and policy agenda that engaged every barrier impeding this goal. Biggs was aware that a reductionist focus on the detection and eradication of pathogenic bacteria was not the solution to all the problems associated with diphtheria and other diseases. Yet he and his successors held firmly to the belief that once the technical means for the prevention and eradication were available, efforts to control a disease such as diphtheria would be maintained.

This belief was misguided. The questions of long-term sustainability of immunization programs, public access to immunization, and long-term public financing of vaccine production and distribution were not addressed.[10] "Public health is purchasable" was the slogan that characterized the vision of Biggs and his colleagues. The slogan expressed the beliefs of a generation of public health reformers that had witnessed what was for them a dramatic decrease in infectious diseases due in part to their own efforts.

From the vantage of the present, the kind of leadership in public health that Biggs and Park exemplified and the powerful role that the institution they guided played in disease control were relatively short lived. Yet they left an important legacy that has largely faded from view. They recognized that infectious disease poses extraordinary challenges to public health in a multi-ethnic, class-stratified country such as the United States. In their minds, the control of these diseases placed an equally extraordinary obligation on the state to prevent such diseases by all available means, in all segments of the population. In an era when there are increasing calls for the national government to turn over the implementation of public programs to private interests—a time, indeed, when the notion of any central role for government is suspect—it is important to remember that the public health triumphs described here could not have been realized without a serious commitment of resources and leadership by those in the public sector.

NOTES

Introduction

1. See Michael Specter, "Russia Fights a Rising Tide of Infection," *New York Times*, 2 October 1994, 10.

2. "Russian Chaos Breeds Diphtheria Outbreak," *Science*, 267 (10 March 1995): 1416–17.

3. "Infectious Diseases Are Found Spreading in Eastern Europe," *Boston Globe*, 8 March 1996, 14.

4. See CDC editorial note following the article "Diphtheria Acquired by U.S. Citizens in the Russian Federation and Ukraine—1994," *Journal of the American Medical Association*, 273 (1995): 1252.

5. T. Popovic, M. Wharton, et al., "Are We Ready for Diphtheria?" 765–67.

6. Ibid. In part this is because through the widespread vaccination of children, the diphtheria bacilli were almost wiped out in most Western countries, so that by the 1980s few adults had enough exposure to the bacilli to boost their natural immunity.

7. This view was expressed in a 1974 editorial in *The Lancet*, "Infectious Diseases—End of a Specialty," as quoted by Dixon in "Infectious Diseases—A 'Dead' Specialty Revives," 779.

8. Here, following Charles Rosenberg, I use diphtheria as a social actor. Rosen-

berg argues that disease concepts "serve in some sense as independent actors, constraining the options of human actors in social situations," and therefore can serve as "a multidimensional sampling device for the scholar concerned with the relation between social thought and social structure." He continues, "The biological character of particular ills defines both public health policies and therapeutic options. Acute and chronic ills obviously confront physicians, governments, and medical institutions with very different challenges, but acute infections themselves vary, for example, in their modes of transmission and thus have very different social connotations" ("Framing Disease: Illness, Society and History," unpublished paper, 1990). Also discussed in his "Introduction: Framing Disease: Illness, Society, and History."

Of course the notion of the "microbe as actor" is most often ascribed to Bruno Latour as discussed in his book *The Pasteurization of France*. My use of diphtheria as an actor differs from his in a number of ways. First, while I argue that diphtheria structures social relations, I do not ascribe to it purpose or intentionality. Second, I try to make the relationship between all the actors in this story—diphtheria, physicians, public health experts, and bacteriologists—stand in some sort of symmetrical relationship to one another through my discussion of the contestations over bacteriological knowledge that emerged in the history of the control of diphtheria. For a fuller discussion of the question of symmetry in Latour's work, see Simon Schaffer, "The Eighteenth Brumaire of Bruno Latour," 175–92.

9. I use the term *interests* here to refer to the historiography of social interests, which looks at the actual interests—professional, factional, and political—of the figures involved in the control of diphtheria. I do not take these interests as historical givens, nor do I argue that diphtheria was controlled only by an appeal to the interests of medical and public health professionals. My use of interest theory follows that proposed by Jardine in "The Laboratory Revolution in Medicine," 304–23.

10. I use the term *identity* of diphtheria to refer to the ways in which infectious disease became defined by the laboratory. The laboratory transformed diphtheria from a symptom-based entity to a germ-based entity. As a result, the laboratory demonstrated that between one-quarter and one-third of cases of "clinical" diphtheria did not have the diphtheria bacilli present. On the other hand, 80 percent of cases of "membranous croup" proved on culture to be caused by diphtheria bacilli. The laboratory then redefined what was known as "true" and "false" diphtheria. See William Park and Alfred Beebe, "Report on Bacteriological Investigations and Diagnosis of Diphtheria, from May 4, 1893, to May 4, 1894." For a discussion of the identity of disease, see also Andrew Cunningham, "Transforming Plague: The Laboratory and the Identity of Infectious Disease," 209–44. The laboratory research on diphtheria was translated into other currency by Biggs: influence, institutions, programs, and ideol-

ogy. This translation is similar to what Bruno Latour describes with respect to Pasteur's success. See Latour, *The Pasteurization of France*.

11. The infection with the bacilli produces a degeneration of cells in the throat that extends to the underlying tissues. The pseudomembrane consists of dead cells, leucocytes (white blood cells), and bacteria (W. Burrows, *Textbook of Microbiology*, 681).

12. Daniel M. Fox, "Social Policy and City Politics," 415–31; and Charles E. Rosenberg, "Making It in Urban Medicine," 163–86.

13. Warner, "The Fall and Rise of Professional Mystery," 111.

14. Peter S. Buck, "Order and Control," 237–67.

15. Starr, *The Social Transformation of American Medicine*, 160, and discussion of this point by Warner in "The Fall and Rise of Professional Mystery," 134–35.

16. Neither of these campaigns has been analyzed by historians. While noting the importance of the antitoxin subscription campaign, for example, most accounts argue that this campaign was evidence of a preexisting public demand for antitoxin. I argue, however, that the campaign itself was orchestrated to serve multiple purposes: to create public demand for the new drug, to undermine potential opposition from physicians, and to promote the power of scientific medicine. For the older view, see John Duffy, *A History of Public Health in New York City, 1866–1960*, 100; William C. Rothstein, *American Physicians in the Nineteenth Century*, 237–77; Wade W. Oliver, *The Man Who Lived for Tomorrow*, 106; C.-E. A. Winslow, *The Life of Hermann Biggs*, 107–30; Terra Ziporyn, *Disease in the Popular American Press*, 42–46.

17. Hermann M. Biggs (1859–1923) was the first director of the Division of Pathology, Bacteriology and Disinfection of the New York City Health Department. Biggs instituted programs for the control of infectious disease based on developments in bacteriology.

18. William H. Park (1863–1939) was the architect of the diphtheria diagnostic and control programs in the New York City Health Department. He was director of the Bureau of Laboratories in the health department.

19. This argument is made by William Park in "The History of the Control of Diphtheria in New York City," 1439–45, and by William Park, M. C. Schroeder, and Abraham Zingher, "The Control of Diphtheria," 23–32.

20. See a discussion of the conflicts between health officials and school medical inspection practices in Alan M. Kraut, "East Side Parents Storm the Schools," 226–56; and John Duffy, *History of Public Health*, 269–70.

21. For a good discussion of the general areas of conflict between physicians and health officers in the Progressive period see John Duffy, *The Sanitarians*, 205–20.

22. Hermann M. Biggs, "Sanitary Science, the Medical Profession, and the Public," 1635–40.

23. This point about the views of Progressive Era reformers is made by David J. Rothman, "The State as Parent," 74.

24. These three types were not identified until 1931 after the period of this study (G. S. Wilson, A. A. Miles, and M. T. Parker, *Topley and Wilson's Principals of Bacteriology, Virology and Immunity*, 2:96).

25. Burrows, *Textbook of Microbiology*, 681.

26. Wilson, Miles, and Parker, *Topley and Wilson*, 3:53–96.

Chapter 1: *The Impossibility of Control*

1. Charles Rosenberg, "Disease in History," 1–15.

2. Frederick W. Andrewes et al., *Diphtheria: Its Bacteriology, Pathology and Immunology*, 30. The term *diphtheria* was introduced by Farr in the Registrar General's Report of 1858, 35.

3. When the diphtheria bacillus lodges on a membrane in the throat, nose, or larynx, the bacilli first multiply in the superficial layers of the mucous membrane. Then they kill the cells, whereupon the blood vessels pour out fluid, which clots and covers the surface, producing a thin, tough layer called a pseudo or false membrane. See B. Davis et al., *Microbiology*, 507; and Harry F. Dowling, *Fighting Infection*, 18.

4. This is the definition of the term *diphthérite* given by F. Loeffler, the codiscoverer of the diphtheria bacillus, in his "The History of Diphtheria," in G. H. Nuttall and G. S. Graham-Smith, *Bacteriology of Diphtheria*, 13–14.

5. In 1855, Bretonneau began to use the name *diphtérie*, which in English was adapted as *diphtheria*. See his "Sur les moyens de prévnir le développement et les progrès de la diphtérie." *Arch. gen. de med.* 5 (1855), 1:6, 256. Citation in Nuttall and Graham-Smith et al., *Bacteriology of Diphtheria*, 441.

6. Erwin Ackerknecht, *Medicine in the Paris Hospital*, 118–19.

7. Knud Faber, *Nosography in Modern Internal Medicine*, 47.

8. Ibid., 34. *Croup*, a Scottish word meaning a cry or hoarse sound, was a ubiquitous term, used at least since the seventeenth century to describe a variety of mild and severe inflammations of the throat, which often involved obstruction of the windpipe.

9. Andrewes et al., *Diphtheria*, 35.

10. See E. H. Ackerknecht, *Rudolph Virchow: Doctor, Statesman, Anthropologist*.

11. Ibid., 16.

12. Faber, *Nosography*, 79.

13. My discussion here follows that of Peter C. English, "Diphtheria and Theories of Infectious Disease," 1–9.

14. Cecilia Mettler, *History of Medicine*, 761.

15. J. Lewis Smith, "Membranous Croup; Diphtheritic Croup; True Croup," 1:438.

16. Ibid.

17. Mettler, *History of Medicine*, 761 (quotes E. Henoch).

18. Andrewes, et al., *Diphtheria*, 37.

19. In 1864, Virchow argued that one had to make a differentiation between the free, superficial exudates in the throat and the diphtheritic forms that were found in the deeper layers and could only be detached by some sort of ulceration of the surrounding tissue. He considered the concept of croup to be clinical, not anatomical. See Rudolf Virchow, "Diphtheria and Croup," and "Commentary from the Proceedings of the Medical Society of Berlin, 14 December 1864." Both papers are found in Rudolf Virchow, *Collected Essays on Public Health and Epidemiology*.

For physicians the connection between croup and malignant sore throat (or diphtheria) had a long history. In the eighteenth century Dr. Francis Home wrote a widely read monograph entitled "An Inquiry into the Nature, Cause, and Cure of Croup" (1765). Home believed that croup was an independent disease caused by exposure to cold. In 1771 Dr. Samuel Bard, a prominent New York City physician in the colonial period and professor of medicine at Kings College, published a monograph arguing that Home's croup was a form of contagious sore throat (also believed to be diphtheria by later authors). See "An Enquiry into the Nature, Cause and Cure of the angina suffocation or, throat distemper, as it is commonly called by the Inhabitants of this City and Colony" (New York: Inslee and Tar, 1771). See also Abraham Jacobi, *A Treatise on Diphtheria*, and Ernest Caulfield, "A History of the Terrible Epidemic Vulgarly Called the Throat Distemper," 219–35.

20. Mettler, *History of Medicine*, 759.

21. Virchow, "Diphtheria and Croup," 485.

22. John Eyler, "William Farr on the Cholera," 80.

23. Abraham Jacobi, *On the Treatment of Diphtheria in America*, 1.

24. The name *zymotic* originated with William Farr and was based on the Greek word meaning "to ferment." Farr's theory of representing the disease process as a form of fermentation, or "zymosis," was an adaptation of Liebig's work on the chemical explanation of putrefaction and fermentation (Eyler, "William Farr," 82). Also see Margaret Pelling, *Cholera, Fever and English Medicine, 1825–1865*.

25. Pelling, *Cholera, Fever and English Medicine*, 98.

26. Ibid., 109.

27. In New York, mortality from diphtheria increased from 1857 to a peak in 1864; it peaked again in 1875, 1881, and 1887. William H. Park and Charles Bolduan, "Mortality," 575.

28. Hirsch, *Handbook of Geographical and Historical Pathology*, 115.

29. William Coleman, *Yellow Fever in the North*, 189. Coleman's argument follows from the discussion by Roger Cooter, "Anticontagionism and History's Medical Record," 87–108.

30. Coleman, *Yellow Fever in the North*, 189.

31. C. E. Billington, *Diphtheria: Its Nature and Treatment*, 162.

32. Joseph E. Winters, *Diphtheria and Its Management*, 621. A portion of Winters's book was published under the same title in *Medical Record*, 28 (5 December 1885): 618–26.

33. Ibid., 623.

34. Joseph E. Winters, "Diphtheria and Its Management: Are Membranous Croup and Diphtheria Distinct Diseases?" 621.

35. Ibid., 622.

36. Ibid., 623.

37. John Harley Warner, *The Therapeutic Perspective: Medical Practice, Knowledge, and Identity in America, 1820–1885*, 1.

38. Isaac A. Abt, ed. *Pediatrics*, 108.

39. Ibid. See also Abraham Jacobi, "A History of Pediatrics in New York," 5–6. For a full history of O'Dwyer's work see his *Intubation in Croup and Other Acute and Chronic Forms of Stenosis of the Larynx*, and W. P. Northrup, *Medical Record*, 65 (1904): 561–64.

40. Jacobi, *On the Treatment of Diphtheria in America*, 3.

41. Augustus Caillé, "A Method of Prophylaxis in Diphtheria," 184.

42. Ibid.

43. Victor C. Vaughan, *Epidemiology and Public Health*, 3.

44. Ibid., 5.

45. Death notices, *New York Times*, 11 June 1883.

46. Regina Markell Morantz, "Feminism, Professionalism and Germs," 470 n. 26.

47. Mary Putnam Jacobi to Mrs. Curtis, 4 September 1883, Box 1, Folder 13, Mary Putnam Jacobi Collection, Schlesinger Library, Radcliffe College.

48. Rhoda Truax, *The Doctors Jacobi*, 198.

49. Jacobi, *On the Treatment of Diphtheria in America*, 3.

50. Ibid., 4.

51. Abraham Jacobi, "Diphtheria Spread by Adults," 344–47.

52. Ibid., 345 (emphasis Jacobi).

53. Ibid. As an example, he noted that tonsillar diphtheria was not very dangerous, while laryngeal and nasal diphtheria were always very serious, with children often dying of strangulation or sepsis.

54. Ibid.

55. Ibid.

56. Ibid.

57. Ibid., 346.

58. Truax, *The Doctors Jacobi*, 198.

59. Jacobi, "Diphtheria Spread by Adults," 346.

60. Ibid., 347.

61. Dr. D. S. Kellogg, "Diphtheria Spread by Adults," Correspondence. *Boston Medical and Surgical Journal*, 23 October 1884, 405.

62. Ibid.

63. Abraham Jacobi, "Follicular Amygdalitis," 593–600.

64. B. Fraenkel, "Angina Lacunaris and Diphtheritic," *Berlin Clinical Weekly*, Nos. 17 and 18, 1886 (as noted by Jacobi in the article "Follicular Amygdalitis," 593).

65. Jacobi, "Follicular Amygdalitis," 593.

66. Ibid., 593–96.

67. Ibid., 612.

68. J. Lewis Smith, "Membranous Croup," 435.

69. Abraham Jacobi, "Diphtheria," 696–97.

70. "Report of the Committee on the Cause and Prevention of Diphtheria," 97.

71. Richard Skolnik, "The Crystallization of Reform in New York City, 1890–1917" (Ph.D. diss., Yale University, 1964), 1–13.

72. *New York Times*, 18 January 1895, reprinted in Allan Schoener, *Portal to America*, 212.

73. John S. Billings, "Municipal Sanitation in New York City," 347.

74. Kate H. Claghorn, "The Foreign Immigrant in New York City," *Report of the Industrial Commission*, 467.

75. Skolnik, "Crystallization of Reform," 14.

76. Jacob Riis, *How the Other Half Lives*, 233.

77. John S. Billings, *Vital Statistics of New York and Brooklyn*, 81.

78. David Ward, *Poverty, Ethnicity, and the American City, 1890–1925*, 61.

79. Riis, *How the Other Half Lives*.

80. The leading causes of death were as follows: consumption (32,781), diarrheal diseases (26,514), pneumonia (24,090), diseases of the nervous system (20,247), stillborn (18,862), diphtheria and croup (15,199). Due to the confusion in reporting cases of diphtheria and croup, data in the report gave the death rates for the two diseases together (Billings, *Vital Statistics*, 36). Since the reporting of contagious disease was by no means a reliable process in the period covered by this report, these data do not give an accurate picture of the prevalence of contagious disease in the city.

81. Billington, *Diphtheria*, 37.

82. Fourteenth Ward: 277.19% and 235.48%; Tenth Ward: 202.06% and 171.28%.

These data are mortality rates per 100,000 of population for two sanitary districts in Wards 14 and 10 (see Billings, *Vital Statistics*, 83–163).

83. The commentary here by contemporary observers is very general. I was not able to find any specific reference to disparities in diphtheria mortality between various groups.

84. Modern studies include Deborah Dwork, "Health Conditions of Immigrant Jews on the Lower East Side of New York: 1880–1914," 27, and Gretchen A. Condran, "Changing Patterns of Epidemic Disease in New York City," 26–41, esp. 40.

85. Claghorn, "The Foreign Immigrant," 474.

86. Report of the New York State Tenement House Committee, 19–21. Hereafter Tenement House Report.

87. See David Ward, "The Internal and Spatial Structure of Immigrant Residential Districts in the Late Nineteenth Century," 357–73, and "The Internal Spatial Differentiation of Immigrant Residential Districts," 335–43.

88. Jacobi, *On the Treatment of Diphtheria in America*, 4.

89. "The Contagious Disease Hospitals of New York City," *Monthly Bulletin of the Department of Health of the City of New York*, 3 (9 September 1913): 191.

90. "Three City Hospitals," *New York Times*, 9 April 1892, 2:2.

91. "A Diphtheria Hospital," *New York Times*, 10 April 1887, 10:2.

92. Ibid.

93. Jacobi, in fact, long advocated for several contagious disease hospitals located around the city, including special ones for middle- and upper-class patients. See his *On the Treatment of Diphtheria in America*, 4.

94. *Annual Report of the Board of Health of the Health Department of the City of New York, for the year ending December 31, 1890* (New York: Martin B. Brown, Printer, 1891), 29. Hereafter *Annual Report Board of Health*.

95. *Tenement House Report*, 20.

96. Phyllis H. Williams, "The South Italians: Health and Hospitals," 470.

97. Annie S. Daniel, "The Wreck of the Home," 628–29.

98. Abraham Jacobi, "Report on the Prevention of Contagious Diseases," in *Collected Essays of Abraham Jacobi*, 152.

99. Ibid., 618.

100. Winters, "Diphtheria and Its Management," 618.

Chapter 2: *The Promise of Control*

1. See Phyllis A. Richmond, "American Attitudes Toward the Germ Theory of Disease (1860–1880)," 428–54.

2. Andrew Cunningham, "Transforming Plague," 224.

3. Ibid. Of course the most well known example of the transformation of disease entities produced by the laboratory is that of tuberculosis. With the discovery of the tubercle bacillus, what had been called phthisis became only one form of tuberculosis. The term *tuberculosis* is now restricted to those manifestations of the disease associated with the bacillus.

4. I use *discourse* here in the Foucauldian sense of the language and practices that shape the objects of which they speak. The concept of discourse helps to explain how diphtheria became a disease that observers believed could be controlled at the very moment when new meanings about the control of disease were being constructed. See also Francois Delaporte, "The History of Medicine According to Foucault," 137–49. I also draw from William H. Park's discussion of the history of the control of diphtheria. Park defined three measures that constituted the control of diphtheria: the bacteriological diagnosis program in 1892, the introduction of diphtheria antitoxin, and finally, the use of toxin-antitoxin for immunization. See W. H. Park, M. C. Schroeder, and A. Zingher, "The Control of Diphtheria," 23–32.

5. See Joan Fujimura, "Crafting Science: Standardized Packages, Boundary Objects, and 'Translation,'" 169–70. The concept of *boundary objects* was developed by Susan Leigh Starr, in S. L. Starr and J. R. Griesemer, "Institutional Ecology, 'Translations,' and Boundary Objects," 387–420. My use of the term *social worlds* follows that of Fujimura, meaning activities and processes that are similar to disciplines but with more fluid boundaries. This is particularly important here because bacteriology as a discipline in the United States was still in its formative period, and Park and Biggs in particular occupied positions as bacteriologists, public health experts, and physicians.

6. Stanley J. Reiser, *Medicine in the Reign of Technology*, 140.

7. Satterhwaite would later author the first American laboratory manual in bacteriology, *Introduction to Practical Bacteriology* (1887), cited in Edward T. Morman, "Clinical Pathology in America, 1865–1915," 199.

8. *Annual Report Board of Health*, 1874–75, 658–83.

9. The identification of the bacillus is credited to Klebs, and in the literature it is often referred to as the Klebs-Loeffler bacillus.

10. This discussion of Loeffler's work is drawn from the chapter on the history of diphtheria in Andrewes et al., *Diphtheria*, 48–55, and Loeffler's discussion of his work in his chapter on the history of diphtheria in Nuttall and Graham-Smith, eds., *The Bacteriology of Diphtheria*, 28–37. His initial paper on diphtheria is "Untersuchungen uber die Bedeutung der Mikroorganismen für die Entstehung der Diphtherie beim Menschen, bei der Taube und beim Kalbe," *Mithh. a. d. k. Gsndhtsamte*, 2 (1884): 421–99.

11. The pure culture method was a difficult one. Careful attention had to be given to controlling the temperature and the chemical content of the culture medium. Failures in growing colonies of bacteria due to the composition or handling of the culture medium were common. Loeffler grew colonies of the bacilli in a medium consisting of three parts calf or sheep serum with one part neutralized veal bouillon, to which had been added 1% peptone, 1% of glucose, and 0.5% of common salt. Andrewes notes that this serum, known as Loeffler's serum, was used exclusively by Loeffler in all of his research because it yielded abundant and luxuriant growths of the bacilli (Andrewes et al., *Diphtheria*, 50).

12. For a discussion of the evolution of Koch's postulates see K. Codell Carter, "Koch's Postulates in Relation to the Work of Jacob Henle and Edwin Klebs," 361.

13. Ibid., 358. Carter notes that Koch sometimes denied that bacteria could ever be found in healthy tissues. Carter credits Koch with the discovery of the phenomenon of healthy carriers in 1893. However, given the findings of Loeffler and later Park on healthy carriers in diphtheria, this point should be questioned.

14. Discussion on Diphtheria, December 20, 1884. *Quarterly Bulletin of the Clinical Society of the New York Post-Graduate Medical School and Hospital*, 1 (1885): 1.

15. Abraham Jacobi, Inaugural Address before the New York Academy of Medicine, February 5, 1885, in *Collected Essays of Abraham Jacobi*, 7:157–78.

16. Ibid., 166.

17. Ibid., 163 and 165. Jacobi noted that deaths from zymotic and infectious diseases resembled those produced by poisons. He was not ruling out a role for bacteria, but he questioned the claim that it should be the sole etiological agent. In a footnote in his address he asked, "Would it be so impossible to judge that the bacterium is an accompaniment of a chemical poison and may be present, or absent, according to the changed condition of the poison?"

18. Carter, "Koch's Postulates," 369–74.

19. Joseph McFarland, "The Beginning of Bacteriology in Philadelphia," 151.

20. See for example, Ferdinand Hueppe, *The Methods of Bacteriological Investigation*, trans. Hermann M. Biggs (New York: D. Appleton and Co., 1886).

21. Donald Fleming, *William H. Welch and the Rise of Modern Medicine*.

22. Lillian E. Prudden, *Biographical Sketches and Letters of T. Mitchell Prudden*.

23. Arnold Eggerth, *The History of the Hoagland Laboratory*.

24. Fleming, *William H. Welch*, 74–75. See also Thomas N. Bonner, *American Doctors and German Universities*.

25. Wade W. Oliver, *The Man Who Lived for Tomorrow*, 54.

26. "The skilled pathologist, he asserted, now must know accurately and fully the structure of the body, both gross and microscopic. . . . He must know and must have had practical experience in examining the various alterations of the body due to dis-

ease or to injury, and must know what modern science reveals as to the agencies causing these alterations. For this reason he must have a practical knowledge of germs and
their action in causing disease; of poisons, their nature and how they act; of various
mechanical injuries and their effects; and of the many social conditions and habits
which may lead to disease and death." L. Prudden, *Biographical Sketches*, 73.

27. This is a reference to a report made by Loeffler in 1887 on his continuing work
in diphtheria, in which he had reported on the investigations that supported his earlier claims. However, there was one case where he found bacteria that were morphologically similar to the diphtheria bacilli, yet when inoculated into guinea pigs were
found to be completely nonvirulent. These bacteria he named *pseudo-diphtheria bacilli*, noting only that further research was needed to clarify their role with respect to
diphtheria (ibid., 97).

28. T. Mitchell Prudden, "On the Etiology of Diphtheria: An Experimental
Study," 331.

29. Ibid., 338.

30. Ibid., 339.

31. T. Mitchell Prudden, "Studies on the Etiology of Diphtheria—Second Series," 446.

32. William H. Welch and Alexander C. Abbott, "The Etiology of Diphtheria,"
25–31.

33. F. Loeffler, "Der gegenwartige Stand der Frage nach der Entstehung der
Diphtherie."

34. E. Roux and A. Yersin, "Contribution a l'étude de la diphtérie."

35. Welch and Abbott quoted Loeffler from the paper cited ("Etiology of Diphtheria," 25–26). *Liquet* is defined as "clear, apparent, or evident" in *Black's Law Dictionary.*

36. See comments by A. Jacobi on the views of American physicians on the relationship between geography and disease in "Epidemics," in James J. Walsh, *History
of Medicine in New York*, 99.

37. Welch and Abbott, "Etiology of Diphtheria," 26.

38. It is appropriate here to note that the etiology of "scarlatina" or scarlet fever was
as much in contention in this period as diphtheria. It complicated much research on
diphtheria because outbreaks of the two often occurred together. As Dowling notes,
"Its main features are a sore throat and a bright red rash, which is frequently spread so
evenly that it appears to be painted on the skin . . . the throat symptoms were often
confused with diphtheria and the rash with measles. In 1884, Loeffler found streptococci in patients with scarlet fever. The explanation of this relationship was not found
until the 1920s" (Harry F. Dowling, *Fighting Infection*, 58).

39. Welch and Abbott, "Etiology of Diphtheria," 28.

40. Loeffler, "History of Diphtheria," 34, in Nuttall and Graham-Smith volume.

41. Prudden, "Studies on the Etiology of Diphtheria—Second Series," 450. Prudden was ultimately recognized for the work he did on mixed infections in diphtheria.

42. The details of his life are found in the biography by Wade W. Oliver, *The Man Who Lived for Tomorrow*. No other personal papers of Park have been located.

43. Park's biographer reports that in the course of their conversation, Park and Prudden discussed the relationship of the Loeffler bacillus to diphtheria. Park stated that, based on demonstrations he had seen in Europe, he believed that diphtheria was caused by the Loeffler bacillus. Prudden then offered Park a scholarship.

44. William H. Park, "Diphtheria and Allied Pseudo-membranous Inflammations," 115–22.

45. His position was still an unusual one. A regular reader of *Medical Record* or *New York Medical Journal* in the years 1890 to 1892 would find that the question of the etiology of diphtheria was not a closed one for most physicians.

46. Park, "Diphtheria and Allied Inflammations," 116. Blood serum is the watery portion remaining after blood has clotted. Park used freshly shed calf or sheep blood. After the blood had clotted, the serum was siphoned off and mixed with a solution of beef broth mixed with small amounts of glucose, peptone, and salt. The mixture was boiled and then filtered through absorbent cotton or filter paper. The mixture was then sterilized and allowed to coagulate. This sterilized blood serum was then stored in tin boxes where it could be kept for months. After the swab had been rubbed against the pseudomembrane, it was rubbed against the surface of the solidified blood serum. The tube with the inoculated blood serum was then kept in an incubator at 37° C for twelve hours. At this time the surface of the blood serum was usually found to be covered with colonies of bacilli that could be seen with the naked eye. The bacteria were then placed on a cover glass, stained with a solution of dye developed by Loeffler of alkaline methyl blue, and examined under a microscope. See William H. Park and Alfred Beebe, "Diphtheria and Pseudo-Diphtheria," 386.

47. Welch and Abbott, "Etiology of Diphtheria," 31.

48. Park, "Diphtheria and Allied Pseudo-membranous Inflammations," 141–47.

49. Ibid., 146.

50. Notes on Hermann Biggs (1859–1923), C.-E. A. Winslow Papers, Yale University Archives. Also see C.-E. A. Winslow, *The Life of Hermann M. Biggs*. The personal papers of Biggs have not been located.

51. David A. Blancher, "Workshops of the Bacteriological Revolution: A History of the Laboratories of the New York City Department of Health" (Ph.D. diss., City University of New York, 1979), 24–25.

52. Richard Hofstadter, *The Progressive Movement, 1900–1915*, 1–16.

53. Joseph D. Bryant (1845–1914) was professor of surgery at University and Bellevue Hospital Medical College. He served as commissioner of health in the New York City Health Department from 1877 to 1893 when he resigned to become the personal physician of President Grover Cleveland. He also served as president of the New York Academy of Medicine, 1895–1896. See Walsh, *History of Medicine in New York*, 4:32.

54. Winslow, *Life of Hermann Biggs*, 80.

55. Hermann M. Biggs, "The History of the Recent Outbreak of Epidemic Cholera in New York," 63–72.

56. L. Prudden, *Biographical Sketches*, 289, and Winslow, *Life of Hermann Biggs*, 80.

57. Hueppe, *Methods of Bacteriological Investigation*, 212. He praised the University of Copenhagen, which in 1883 had been the first institution to establish an institute devoted to medical bacteriology in connection with the Botanical Institute. In Germany, Hueppe noted, bacteriology was studied in a variety of institutional settings: in the pathological institute in Munich; in the hygienic institute in Göttingen; and in his own school at Wiesbaden, where hygienic bacteriology was associated with the chemical institute of Fresenius. There was also a laboratory under the Office of the Imperial Board of Health that investigated municipal problems.

58. Ibid., 214.

59. T. Mitchell Prudden, "On Koch's Methods of Studying Bacteria."

60. Ibid., 229.

61. "Drs. Jacobi and Prudden Quit Board of Health," *New York Times*, 25 June 1892, 8:1.

62. *New York Times*, 26 June 1892, 9:7.

63. Smith's remarks and a detailed discussion of this incident is found in Gordon Atkins, "Health Housing and Poverty in New York City, 1865–1898" (Ph.D. diss., Columbia University; Ann Arbor, Mich.: Gordon Atkins, 1947), 252.

64. Ibid., 252.

65. Ibid., 254.

66. *New York Times*, 13 July 1892, 4:2.

67. *New York Times*, 13 August 1892, 8:4. Also see Atkins, "Health, Housing and Poverty," 253, and *New York Medical Journal*, 27 August 1892.

68. T. Mitchell Prudden to William Jay Schiefflin, August 12, 1892. Prudden Papers, Yale University Archives.

69. Letter signed "Ignotus," *New York Medical Journal*, 9 July 1892, 49–50.

70. Blancher, "Workshops," 33.

71. Winslow, *Life of Hermann Biggs*, 93.

72. Charles E. Rosenberg, *The Cholera Years: The United States in 1832, 1849, and 1866*.

73. See Blancher's discussion of the cholera epidemic and its role in the establishment of the bacteriological laboratory ("Workshops," 34–47).

74. George Shrady, editorial, *Medical Record*, 13 August 1892, 193.

75. *New York Times*, 5 September 1892, 1.

76. William H. Welch to Prudden, September 13, 1892. Prudden Papers, Yale University Archives.

77. Charles Wilson, Walter Wyman, and Cyrus Edson, "Safeguards Against the Cholera," 491.

78. *New York Times*, 15 September 1892, 2:3.

79. Biggs, "History of the Recent Outbreak," 63–72, and Edward K. Dunham, "The Bacteriological Examination of the Recent Cases of Epidemic Cholera in the City of New York," 72–80.

80. Wilson et al., "Safeguards," 491.

81. Editorial, *Medical Record*, 10 September 1892, 312. For the view from Hamburg, see Richard Evans, *Death in Hamburg*.

82. *Medical Record*, 14 January 1893, 48–49.

83. *Medical Record*, 9 September 1893, 333.

84. *Medical Record*, 14 January 1893, 49.

85. Winslow, *Life of Hermann Biggs*, 97.

86. Biggs, "History of the Recent Outbreak," 71.

87. Report of the Committee of Conference of the New York Academy of Medicine, *Medical Record*, 29 October 1892, 516. The Committee was asked by the president of the Board of Health to examine the preparations made by the department for the care and treatment of cholera patients and measures to prevent the spread of disease. No explicit mention of bacteriological examinations was made in the published version of the report. Also see Discussion in Section on Public Health, New York Academy of Medicine, *Medical Record*, 24 September 1892, 375–76.

88. *Medical Record*, 24 September 1892, 376.

89. T. Mitchell Prudden, "The Public Health: The Duty of the Nation in Guarding It," 247.

90. Hermann M. Biggs, "The Organization, Equipment and Methods of Work of the Division of Pathology," 303–28.

91. Oliver, *Man Who Lived for Tomorrow*, 75–76.

92. *Annual Report Board of Health*, 1893, 10.

93. Ibid., 64

94. Editorial, *Medical Record*, 4 February 1893, 144–45.

95. William H. Park, "Diphtheria and Other Pseudo-membranous Inflammations," 162.

96. *Annual Report Board of Health*, 1893, 67.

97. Editorial, *Medical Record*, 1 July 1892, 17.

98. Editorial, *New York Medical Journal*, 58 (July 1893): 17.

99. Park, "Diphtheria and Other Pseudo-membranous Inflammations," 163.

100. Dr. Stuyvesant F. Morris, Letter to the Editor, *New York Medical Journal*, 27 November 1893, 666.

101. Louis C. Acer, "The First Four Months of Diphtheria Cultures in Brooklyn," 93.

102. "Bacteriology and the Prevention of Diphtheria," *British Medical Journal*, 10 November 1894, 1067–68.

103. See Rosenberg, "Making It in Urban Medicine," 163–86.

104. Minutes of the Meeting of the Practitioner's Society of New York, *Medical Record*, 25 November 1893, 695.

105. Ibid., 697.

106. Hermann Biggs, "Preventive Medicine in the City of New York," 637.

107. Ibid.

108. Ibid., 697.

109. "Bacteriological Diagnosis," Letter to the Editor, *Medical Record*, 23 November 1894.

110. Barbara G. Rosenkrantz, "Cart before Horse," 61.

111. Park, "Diphtheria and Other Pseudo-membranous Inflammations," 164.

112. *Annual Report Board of Health*, 1893, 71. Biggs refers here to taking inoculations. Park's kit consisted of two tubes—one containing the swab that was rubbed across the throat, the other containing solidified blood serum. The physician had to rub the swab across the throat or any visible membrane, then rub it across the surface of the blood serum in the second tube. This is what is referred to as an inoculation.

113. William H. Park and Alfred Beebe, "Report on Investigations and Diagnosis of Diphtheria, from May 4, 1893 to May 4, 1894."

114. Walter F. Chappell, "Vexed Questions in the Bacteriology of Diphtheria," 457.

115. Park, "Diphtheria and Other Pseudo-membranous Inflammations," 162.

116. Park and Beebe, "Report on Investigations."

117. Ibid., 28.

118. Ibid., 53.

Chapter 3: *The Selling of the Antitoxin*

1. A number of historians have noted the importance of the newspaper campaigns to the introduction of antitoxin in New York City. John Duffy points to the good relations between the health department and the newspapers that inspired the

Herald to initiate its subscription drive. See Duffy, *A History of Public Health*, 100. Rothstein's interpretation is that public interest pushed the *Herald* to launch its campaign. See William G. Rothstein, *American Physicians in the Nineteenth Century*, 273–77. Wade Oliver, Park's biographer, noted that the *Herald* campaign came in the wake of the publicity over antitoxin (*The Man Who Lived for Tomorrow*, 106). Winslow, Biggs' biographer, provides a fuller account; yet he too cites the campaigns in both the *New York Herald* and the *New York Times* as evidence of public demand for antitoxin (*Life of Hermann M. Biggs*, 107–30). Terra Ziporyn discusses the introduction of antitoxin as the news event in the 1890s that garnered more coverage in the newspapers and magazines between 1890 and 1920 than any other aspect of the disease. Her discussion does not, however, connect this media coverage to ongoing medical or public health interest in diphtheria. See T. Ziporyn, *Disease in the Popular American Press*, 42–46. Jean Howson's unpublished paper on the newspaper coverage of the introduction of antitoxin to New York City provides the most complete account and analysis of *New York Times* and *Herald* coverage. Her thesis is that the *Herald* campaign was not a newspaper gimmick, but rather it was part of a calculated public relations strategy adopted by the health department to gain financial support from the city for the production of antitoxin. See Howson, "Sure Cure for Diphtheria," 27. Thanks to Bert Hansen for bringing this paper to my attention.

2. Historians' treatment of the subscription campaign organized by the *New York Herald* (hereafter the *Herald*) generally follows the line of argument put forth by William G. Rothstein that "popular demand for adoption of the anti-toxin put pressure on government public health authorities, who in turn were able to induce physicians to use the therapy." He further notes that "parents with children dying from diphtheria were eager to accept a therapy which had been recommended by so many illustrious physicians and public health authorities." While the influence of these parents is undoubtedly true, this argument implies that public demand for the antitoxin preceded and pushed physicians and public health experts to accept the new remedy (*American Physicians in the Nineteenth Century*, 273–277). See also John Duffy, *A History of Public Health*, 100; and Russell C. Maulitz, "Physician vs. Bacteriologist," 96.

3. See Nicolas Jardine, "The Laboratory Revolution in Medicine as Rhetorical and Aesthetic Accomplishment," 304–23.

4. For a discussion of these campaigns see Paul Weindling, "From Medical Research to Clinical Practice," 78–79.

5. E. Roux and A. Yersin, "Contribution a l'étude de la diphtérie," 629.

6. C. Fraenkel, "II. Immunisierungsversuche bei Diphtherie," *Berlin klinische Wochenschrift* 27 (1890): 123. Discussion of this paper and Behring and Kitasato is taken from Arthur Bloomfield, *A Bibliography of Internal Medicine*, 257–59.

7. E. Behring and S. Kitasato, "Über das Zustandekommen der Diphtherie-Immunität und der Tetanus-Immunität bei Thieren," *Deutsche med. Wchnsch.* 16 (1890): 113–14.

8. E. Behring, "Untersuchungen über das Zustandekommen der Diphtherie-Immunität bei Thieren," *Deutsche med. Wchnsch.* 16 (1890): 1145–48. He concluded that the blood of immune rabbits possessed toxin-destroying properties. These properties were also present in the cell-free serum obtained from such animals. The properties were so lasting that they remained effective when injected into other animals. The toxin-destroying properties were not present in the blood of animals not immune to diphtheria. See Hubert A. Lechevalier and M. Solotorovsky, *Three Centuries of Microbiology*, 218–23 and n. 14.

9. Bulloch cites this story as true; however, H. Schadewaldt, the author of the entry on Behring in the *Dictionary of Scientific Biography* claims that it is unlikely that sufficient serum was available at that time. See William Bulloch, *The History of Bacteriology*, 261, and Behring, s.v. in *Dictionary of Scientific Biography*, 1:574–78.

10. H. Kossel, "Ueber die Blutserumtherapie bei Diphtherie," *Deutsche med. Wchnsch.* 20 (1894): 823 (as discussed in Lechevalier and Solotorovsky, *Three Centuries of Microbiology*, 224).

11. E. Roux and L. Martin, "Contribution a l'étude de la diphthérie (serum therapie)," 609.

12. E. Roux, L. Martin, and A. Chaillou, "Trois cents cas de diphthérie traites par le serum antidiphthérique," 640.

13. V. C. Vaughan, "Immunization Against Diphtheria by the Employment of a Toxin-Antitoxin Mixture: A Caution," *Journal of Laboratory and Clinical Medicine* 5 (February 1920): 334–35 (as recounted in H. Dowling, *Fighting Infection*, 37).

14. Lechevalier and Solotorovsky, *Three Centuries of Microbiology*, 224.

15. Winslow, *Life of Hermann Biggs*, 111.

16. *New York Times*, 27 August 1894, 1.

17. Ibid.

18. *New York Times*, 24 August 1894.

19. Quoted by Biggs in "Antitoxine in New York," *New York Times*, 2 December 1894.

20. Ibid.

21. See Winslow, *Life of Hermann Biggs*, 112, and L. Prudden, *Biographical Sketches*, 45.

22. "The Prevention of Diphtheria," *Journal of the American Medical Association*, 3 (November 1894): 698–99.

23. Ibid., 698.

24. Ibid., 699.

25. Jean E. Howson's excellent paper discusses both the *Herald* and *New York Times* coverage. There are problems with her analysis because she did not consult health department records or other medical journal articles about diphtheria in the fall of 1894 and was therefore unaware of ongoing debates on the disease in the medical community. Also see Ziporyn, *Disease in the Popular Press.*

26. Editorial, "The New Foe of Diphtheria," *New York Times*, 6 September 1894, 4.

27. *New York Times*, 15 October 1894.

28. *New York Times*, 17 October 1894.

29. "Diphtheria's New Treatment" (reprinted from *London Times*), *New York Times*, 9 November 1894.

30. Ibid.

31. Ibid.

32. Ibid.

33. "The Serum Remedy in New York," *New York Times*, 21 November 1894.

34. "Anti-toxine Distributed," *New York Times*, 26 November 1894.

35. George F. Shrady, *Medical Record*, 10 November 1894.

36. See Jonathan Liebenau, "Public Health and the Production of and Use of Diphtheria Antitoxin in Philadelphia," 223. Liebenau quotes Oliver, *The Man Who Lived for Tomorrow*, 105. However, Oliver is careful to note that Park's production of antitoxin was "the first time in the world by sanitary authorities," 105. David Blancher argued that the health department "inaugurated the first systematic production of antitoxin in this country" ("Workshops," 112).

37. See unpublished paper by Jon M. Harkness, "Rabies and Research: A History of the New York Pasteur Institute," presented at the American Association for the History of Medicine Annual meeting, April 1989. I am grateful to Bert Hansen for bringing this paper to my attention and for pointing out the New York Pasteur Institute's role in diphtheria antitoxin production in New York City.

38. Paul Gibier, "The Treatment of Diphtheria with the Antitoxine Made at the New York Pasteur Institute," 1.

39. Ibid., 4. The lack of attention to Gibier's serum is particularly striking given that the firm of Lehn & Fink was also the distributor of his product.

40. "Spurious Antitoxine Sold," *New York Times*, 13 December 1894.

41. Blancher, "Workshops," 109.

42. "Antitoxine in New York," *New York Times*, 2 December 1894.

43. Ibid.

44. Ibid.

45. The biographers of Biggs, Prudden, and Park all agree that horses were purchased by Biggs and Prudden in September 1894. Prudden apparently paid for four

horses and the removal of 80 tons of coal at the stables owned by the College of Physicians and Surgeons where the production of antitoxin began (Prudden, *Biographical Sketches*, 45).

46. Editorial, *New York Times*, 3 December 1894, 4.

47. "Yonkers Epidemic Abating," *New York Times*, 2 December 1894.

48. *New York Times*, 14 December 1894.

49. In 1894, the *New York Herald*, owned by James Gordon Bennett Jr., the son of the founder of the paper, was one of the more widely read newspapers in the city with a circulation much higher than the *New York Times*. See Schudson, *Discovering the News*, 111.

50. "Right Treatment of Diphtheria," *New York Herald*, 25 November 1894, 3.

51. "Anti-Toxine for the Poor," *New York Herald*, 10 December 1894.

52. Ibid.

53. Shrady was the *Herald*'s medical editor and also a consulting physician to the Willard Parker Hospital and editor of the influential journal *Medical Record*.

54. "Anti-Toxine for the Poor."

55. Ibid.

56. *New York Herald*, 11 December 1894.

57. Charles Rosenberg, *The Care of Strangers*, 173.

58. "All Eager for Anti-toxine," *New York Herald*, 11 December 1894.

59. Ibid.

60. "Anti-toxine Grows in Favor," *New York Herald*, 13 December 1894.

61. "Doctors Discuss Anti-Toxine," *New York Herald*, 14 December 1894.

62. *New York Herald*, 13 December 1894.

63. *New York Herald*, 14 December 1894.

64. Ibid.

65. "Cultivating Anti-Toxine," *New York Herald*, 15 December 1894.

66. "Doctors Praise *The Herald*'s Efforts," *New York Herald*, 16 December 1898.

67. "Life Saved by Anti-toxine," *New York Herald*, 23 December 1894.

68. Paul Gibier, "Further Notes Upon the Treatment of Diphtheria," 29–38. Gibier claimed that the New York Pasteur Institute (NYPI) sent out over 11,000 doses of diphtheria antitoxin between September 1894 and November 1895. It is not possible to determine from his report how many doses were sold before January 1895.

69. In June 1895 he acknowledged that his earlier serum had not attained the maximum antitoxin power in December 1894. Gibier provided no details in his publications of the strength of his antitoxin over the course of the next months (*New York Therapeutic Review* 3 [December 1895]: 93). By 1895 advertisements ran in the *New York Therapeutic Review* stating that the NYPI had produced the first antitoxin in the United States.

70. "Only Imported Antitoxin Effective," *Medical Record*, 29 December 1894, 815.

71. "Experts on Anti-Toxine," *New York Herald*, 18 December 1992.

72. Howson, "Sure Cure for Diphtheria," 29.

73. Edson quoted in "Doctors Discuss Antitoxine," *New York Herald*, 14 December 1894, 4. The actual number of horses under treatment by the health department by mid-December is difficult to gauge. Some reports mention fifteen; others claim that thirty or forty horses were under treatment at this point. See H. Biggs, "Some Experiences in the Production and Use of Diphtheria Antitoxin," *Medical Record*, 20 April 1895, 481.

74. Edson's figures did not add up. According to him, one horse could produce enough serum to treat fifteen cases of diphtheria per month. Fifteen horses then would provide enough serum to treat approximately 225 cases per month. But with an average of 200 cases per week, the health department obviously did not have enough antitoxin to meet the needs of the city. It could be that he was misquoted. The following year Biggs claimed that thirty to forty horses were under treatment in the health department by mid-December. See sources in note 73 above.

75. "Will Make Antitoxine," *New York Times*, 21 December 1894. Flower's request included $4,000 for the laboratory, $23,000 for salaries and other expenses, and $25,000 for horses and stables.

76. Ibid.

77. Ibid.

78. For a contemporary account, see Simon Sterne, "Mayor Strong's Experiment in New York City," 539–53. Also see Gordon Atkins, *Health, Housing, and Poverty*, 231–58.

79. Minutes of the New York City Board of Health, 6 January 1894, 93.

Chapter 4: *The Death of Bertha Valentine*

1. Outrage over deaths from the use of antitoxin also brought out physicians' concerns about the drug in Germany. See discussion of the death of the son of eminent German physician Paul Langerhans in Paul Weindling, "From Medical Research to Clinical Practice," 80.

2. See John V. Pickstone's introduction to his edited volume, *Medical Innovations in Historical Perspective*, 5.

3. The most thorough and important source for the history of medical therapeutics in the United States is John Harley Warner, *The Therapeutic Perspective*, 1986.

4. *Medical Record*, 12 January 1895, 50.

5. Editorial, *New York Times*, 1 April 1895. The editorial also misrepresents the

European situation, where discussions of the side effects associated with antitoxin and disappointing results of its use in clinics had been reported. Weindling, "From Medical Research," 80.

6. Editorial, *New York Times*, 3 April 1895.

7. Ibid., 8.

8. "Value of Antitoxine," *New York Times*, 5 April 1895.

9. Hermann Biggs, "Some Experiences in the Production and Use of Diphtheria Antitoxin," 481.

10. Ibid., 484.

11. The strength of the antitoxin produced in the New York City laboratory was determined according to the system designed by Behring. There were three Behring solutions available, which varied in strength. Park and Williams were in the process of refining the methods used by Behring and by Roux in France in order to produce serum of more consistent strength. William H. Park and Anna W. Williams, "The Production of Diphtheria Toxin," 164–85.

12. See Winslow, *Life of Hermann Biggs*, 117. Park based his recommendations on the appropriate dosage on "the extent and intensity of the disease; also, but to a less degree, by the size of the patient and the duration of the illness." He used a single dose of 800 units, 8 c.c. of Behring's standard No. 2 solution for young children; for older children and adults, 1,000 units. To children who were seriously ill or who already showed the effects of the toxin, he gave a dose of 1,500 to 2,000 units of Behring's No. 3. When Park and Williams were able to produce a stronger, more concentrated serum, these dosages changed. See William H. Park, "The Clinical Use of Diphtheria Antitoxin," 256.

13. Joseph Winters, "Diphtheria and Its Management: Are Membranous Croup and Diphtheria Distinct Diseases?" 617–31.

14. "Discussion on the Use of Diphtheria Antitoxin," 204–215.

15. Ibid., 210. Note that the cause of Bertha Valentine's death was not established during this period. It is most likely that she died of anaphylaxis, that is an allergic reaction to the products of the horse serum. This phenomenon would not be defined until 1902.

16. Ibid., 623.

17. See Winslow, *Life of Hermann Biggs*, who characterizes Winters as "one of those pathetic figures who will be remembered in the history of American medicine as a protagonist of error and a foil for the open-minded and clear-thinking medical men of his day and generation" (129).

18. Editorial, *New York Medical Journal*, 13 April 1895.

19. Park, "Clinical Use of Diphtheria Antitoxin," 255.

20. Ibid., 256.

21. Ibid.

22. "Discussion on Diphtheria Antitoxin," 645–49.

23. William Park, "The History of Diphtheria in New York City," 1443. For an excellent discussion of the European situation with respect to hospital statistics on the use of antitoxin in children's hospitals, see Paul Weindling, "From Isolation to Therapy," 124–45.

24. Warner, *Therapeutic Perspective*.

25. "Discussion on Diphtheria Antitoxin," 647.

26. Ibid., 649.

27. Ibid., 648.

28. Ibid., 647.

29. Editorial, "Antitoxin Treatment of Diphtheria," *Archives of Pediatrics*, 12 (June 1895): 446.

30. W. H. Thomson, "How the Facts About the Antitoxin Treatment of Diphtheria Should Be Estimated," 893.

31. Ibid.

32. These men were all prominent members of the New York Academy of Medicine. O'Dwyer was the inventor of the intubation tube (see above, chap. 1).

33. Report of the American Pediatric Society Collective Investigation into the Use of Antitoxin in the Treatment of Diphtheria in Private Practice, 481–503.

34. Ibid., 482

35. Biggs provided the data from Chicago, where the antitoxin used was produced by the New York City Health Department.

36. Report of the American Pediatric Society, 483.

37. Ibid., 501.

38. Ibid., 503.

39. The problem for physicians was, of course, potential errors in the bacteriological report, but also the time delay in receiving a report. As L. Emmet Holt noted, "When a physician makes a culture he waits for the bacteriological diagnosis. He is at last positive of this diagnosis, but loses his patient. He should trust to clinical symptoms, and not wait for a bacteriological diagnosis" ("Discussion on Antitoxin," 70).

40. Report of the American Pediatric Society, 532.

41. Editorial, *Medical Record*, 4 July 1896.

42. He continued: "I have changed my ways in the last fifteen years and have learned much, but have not yet learned, or am not quite positive that I have to exclude from the diagnosis of diphtheria, those cases that are of streptococcus origin. The worst cases are not those of the simple bacilli alone, but those of mixed form" (Jacobi, "Discussion on Antitoxin," 536)

43. See Rothstein, *American Physicians in the Nineteenth Century*, 281.

44. Warner, *Therapeutic Perspective*, 279–83.

45. The editors of *Medical News* wrote, "It is well for scientific medicine, especially for therapeutics, [that] the conservative element in the profession is so strong. It saves many a premature acceptance of what seems at first a useful remedy. What good grounds there were for a reluctant conservatism in the acceptance of diphtheria serum as remedy, may be judged from the fact that its first much-vaunted successes were really not due to the antitoxin at all, but to expectant treatment, for the serums first employed were of such low value (fifty antitoxin units, and even lower) that whoever should pretend now to base statistics on the use of corresponding amounts would be laughed to scorn" ("Serum-Therapy in Diphtheria," 112).

Chapter 5: *The Promise Unfulfilled*

1. An important discussion of this point is Judith W. Leavitt's "'Typhoid Mary' Strikes Back," 608–29. The following studies address some of the issues related to the shifts in public health practice as a result of the introduction of bacteriology: Lloyd Stevenson, "Science Down the Drain," 1–26; Howard D. Kramer, "The Germ Theory and the Early Public Health Program in the United States," 233–47; and Barbara G. Rosenkrantz, "Cart Before Horse" 55–73.

2. The properties of the diphtheria bacilli (i.e., morphological appearance, appearance after staining, growth on bouillon, and virulence when inoculated into guinea pigs) was being questioned during this time. In particular there was still discussion about the pseudodiphtheria bacillus and its relation to the Klebs-Loeffler bacillus. See Myer Solis-Cohen, "Latent Diphtheria," 30–37, and Anna W. Williams, "Persistence of Varieties of the Bacillus Diphtheriae and of Diphtheria-like Bacilli," 83–108.

3. Monomorphism asserted the impossibility of any physiological or morphological variation within a bacterial species, so that the individual organisms of a given species shared a characteristic form, grew in the same manner on specific media, and caused definite disease without undergoing any changes. See Olga Amsterdamska, "Medical and Biological Constraints," 657–87.

4. See a recent discussion of this question with respect to typhoid in J. Andrew Mendelsohn, "'Typhoid Mary' Strikes Again," 268–77; and the letter to the editor by Judith W. Leavitt in *Isis*, 86, no. 4 (1995): 617–18. While both authors raise some important points about carriers and public health policy, my discussion of diphtheria carriers addresses some areas overlooked in both accounts.

5. Leavitt, "'Typhoid Mary' Strikes Back."

6. "Regulations Regarding the Isolation of Cases of Diphtheria in Private Houses

by Health Department," and "Circular of Information Regarding the Danger of Cases of Diphtheria Communicating the Disease to Others after Convalescence," *Medical Record,* 21 March 1896, 431–32.

7. Editorial, "The Isolation of Diphtheria," *Archives of Pediatrics,* May 1896, 431–32.

8. Ibid., 432.

9. William Park and Alfred Beebe, "Report on Bacteriological Investigations and Diagnosis of Diphtheria," 50.

10. Ibid., 51.

11. Ibid., 41.

12. Ibid. After the publication of the 1894 paper, Park gave other examples of diphtheria outbreaks that he had traced to unrecognized mild or convalescent cases that had ultimately convinced him of the danger from such cases. See Park, "Some Recent Studies on the Communicability and Treatment of Diphtheria and Pseudo-diphtheria," 340–63.

13. Ibid., 51.

14. Park, "Some Recent Studies." Biggs's comments appear in the discussion of Park's paper, 360.

15. H. Biggs, "Some Investigations as to the Virulence of the Diphtheria Bacilli Occasionally Found in the Throat-Secretions in Cases Presenting the Clinical Features of Simple Acute Angina," 249.

16. Ibid., 279.

17. Virulence was defined by Park and associates as the ability of the bacilli to produce toxin. To determine the virulence of bacteria, they were grown in pure culture and then injected into guinea pigs. If the animals died, the bacteria were considered to be virulent for humans as well.

18. Biggs, "Some Investigations," 281.

19. Ibid.

20. Maulitz notes that Meltzer was somewhat of a critic of reductionist bacteriology: "His argument rested on a demonstration of the complexity of the host-parasite relationship, one which he felt bacteriology tended to oversimplify." See Russell Maulitz, "Physician versus Bacteriologist," 100–101 and 106 n. 28.

21. Discussion, *Transactions of the Association of American Physicians,* 11 (1896): 259.

22. Ibid.

23. Ibid., 260.

24. Ibid., 261.

25. Hibbert W. Hill, "The Official Definition of Diphtheria," 243.

26. Ibid., 245.

27. Autobiography of Anna Wessels Williams, chap. 14, p. 4. MS, Anna Wessels Williams Papers, Schlesinger Library, Radcliffe College.

28. Francis P. Denny, "Prize-winning Essay," 54–61.

29. Barbara Gutmann Rosenkrantz, *Public Health and the State*, 108 n. 21.

30. Denny, "Prize-winning Essay," 63.

31. Park and Beebe examined 330 healthy persons in whom no contact with diphtheria was known and found the presence of virulent bacilli in eight cases, two of which later developed diphtheria. Using their number of 1%, in a population of 1,809,203 in the city in that year there could be as many as 18,000 such persons in the population.

32. Rosenkrantz, *Public Health and the State*, 66.

33. "Preliminary Report of the Committee on Diphtheria Bacilli in Well Persons," 1–15.

34. "Report of Committee on Diphtheria Bacilli in Well Persons," 74–77.

35. Ibid., 78.

36. Ibid., 82.

37. Ibid., 84.

38. Ibid.

39. Ibid.

40. Ibid., 85.

41. Ibid., 80.

42. J. Ledingham and J. A. Arkwright, *The Carrier Problem in the Infectious Diseases*, 5.

43. Charles F. Craig, "Unrecognized Infections in Production of Carriers of Pathogenic Microorganisms," 828.

44. Henry J. Nichols, *Carriers in Infectious Diseases*, 18.

45. Editorial, "Diphtheria and Boards of Health," *Boston Medical and Surgical Journal*, 6 June 1901, 566.

46. Ibid.

47. E.g., Loeffler (1887); von Hoffman (1888). For a complete survey of this work up until 1900, see Wesbrook, Wilson, and McDaniel, "Varieties of the Bacillus Diphtheriae," 198–223.

48. Amsterdamska, "Medical and Biological Constraints," 663.

49. Williams worked closely with Park on all aspects of the diphtheria work from the time of her appointment to the bacteriological laboratory in 1894. She was coproducer with Park of the most potent antitoxin in the world. See biographical essay by Elizabeth Robinton, "Anna Wessels Williams," 737–39. See Anna Wessels Williams, "Persistence of Varieties of the Bacillus Diphtheriae," 83–108.

50. The question of the relationship between the virulent and nonvirulent strains

of diphtheria bacilli remained an unsolved problem even as late as the 1950s. The use of the terms *nonvirulent* or *avirulent diphtheria bacilli* in itself posed a problem. If the two were so different, and the avirulent strain could not become virulent, then the nomenclature was clearly inappropriate. Though avirulent strains were not seen as a factor in causing diphtheria, researchers still found these avirulent strains in undoubted cases of clinical diphtheria. It was thought that such strains then had some "subtle, unexplained, pathogenic power" or that they seemed to play some role in naturally acquired diphtheria immunity. See Victor J. Freeman, "Studies on the Virulence of Bacteriophage-infected Strains of Corynebacterium Diphtheriae," and M. Frobisher, E. Parsons, and E. Updike, "The Correlation of Laboratory and Clinical Evidence of Virulence of C. Diphtheriae," 543–48.

51. Anna I. von Sholly, "A Contribution to the Statistics on the Presence of Diphtheria Bacilli in Apparently Healthy Throats," 337–46.

52. Ibid., 338.

53. Ibid.

54. See W. Park, A. Williams, and C. Krumweide, *Pathogenic Microörganisms*, 159.

55. Letter from W. C. Woodward to Charles V. Chapin, October 29, 1901, Chapin Papers, John Hay Library, Brown University Library Archives.

56. Charles V. Chapin, "The Restriction of Contagious Diseases in Cities," 994.

57. Charles V. Chapin, *The Sources and Modes of Infection*, 67.

58. W. L. Moss, "Diphtheria Bacillus Carriers," Trans. of the 15th International Congress on Hygiene and Demography (Washington: Government Printing Office, 1913), 4:165.

59. Specifically, he noted, "If 2% of the community in Washington is infected, and an average of 5,577 persons were typically mingling among a population of 273,303 persons for 365 days and as a result of that contact only 675 cases of diphtheria developed . . . this seems like a relatively small number" (Woodward to Chapin, November 5, 1901).

60. Hill to Chapin, March 1, 1910. Brown University Library.

61. W. H. Park, "Some Factors Which Lead to the Increase and Decline of Communicable Diseases Among Men and Animals," 193.

62. See James A. Doull and H. Lara, "The Epidemiological Importance of Diphtheria Carriers," 508–29. This study identified and tested carriers and charted the time elapsed between discovery of the carrier and reporting of cases in order to estimate the number of contacts exposed to the carriers and the number of expected and actual cases due to that exposure. The data were adjusted for age and the seasonal variation in diphtheria incidence in the city.

63. Ibid., 528. However, the authors offered a caveat on this last point. Their study

did not rule out that differences in risk of attack due to carriers could exist based on the differences in the bacilli they harbored.

64. William Park, M. C. Schroeder, and A. Zingher, "The Control of Diphtheria," 24.

65. See Wade H. Frost, "Infection, Immunity and Disease in the Epidemiology of Diphtheria," 325–43; and Wade H. Frost, M. Frobisher Jr., V. A. Van Volkenburgh, and M. L. Levin, "Diphtheria in Baltimore," 568–86.

66. Ledingham and Arkwright, *The Carrier Problem*, 1.

67. Nichols, *Carriers in Infectious Diseases*, 18.

68. Mendelsohn raises this point with respect to typhoid and Weindling argues that some bacteriologically trained health experts in Germany raised this point as well (" 'Typhoid Mary' Strikes Again," 273). See also Paul Weindling, "From Medical Research to Clinical Practice," 81.

69. See a discussion of this point in John Cassel, "The Contribution of the Social Environment to Host Resistance," 107–23.

Chapter 6: *The Use of Active Immunization*

1. Because of World War I and the need to run immunization trials within each country, only preliminary attempts were made to immunize large numbers of children in France, England, and Germany before 1930. Other cities in the United States did not have the necessary personnel or resources to conduct widespread campaigns in the 1920s. See H. J. Parish, *Victory with Vaccines*, 89–94; and John Duffy, *The Sanitarians*, 201.

2. Jane Lewis documents the delays in efforts to immunize schoolchildren in England and Canada during the same period when immunization began in New York City ("The Prevention of Diphtheria in Canada and Britain 1914–1945," 163–76).

3. Wade W. Oliver writes, "Yes there were thousands of young lives that had been spared, but this matter of immunization against diphtheria was not proceeding fast enough for Park. . . . If only a way could be found to determine which children are susceptible to diphtheria and which ones are immune! If such a test were available, the forces of the Health Department could be concentrated upon immunizing the susceptibles. Then, genuine, dramatic progress ought to be made. As if in answer to a prayer, such a test for susceptibility to diphtheria was described by Béla Schick" (*The Man Who Lived for Tomorrow*, 319).

4. Park's textbook, *Pathogenic Microörganisms*, coauthored with Anna Williams, went though eleven editions from 1899 to 1939 and was widely used by medical students and bacteriologists.

5. Ibid., 1442.

6. Ibid. The number given by Park is difficult to evaluate. The Board of Health Report for 1896 lists 1,207 cases immunized with antitoxin. I could not determine how Park arrived at the figure of 25,000 injections. See Annual Report, Board of Health, 1896, 317.

7. Hermann Biggs, "The Serum-Treatment and Its Results," 98. (This was originally a speech read before the New York Academy of Medicine in March 1899.)

8. Ibid., 100.

9. The evidence for his view came from a short trial conducted in 1897, where immunizing doses that were routinely being given in an institution were stopped. Several cases of diphtheria then occurred.

10. Biggs, "The Serum Treatment," 104

11. Ibid., 103.

12. John S. Billings Jr., "Ten Years Experience with Diphtheria Antitoxine," 1311.

13. Ibid., 1312.

14. Discussion of H. W. Hill, "Sources of Error in the Laboratory Diagnosis of Diphtheria," *Public Health Papers and Reports*, 33, part 1 (1907): 226.

15. William Park, M. C. Schroeder, and A. Zingher, "The Control of Diphtheria," 25.

16. S. Josephine Baker, *Fighting for Life*, 56.

17. Francis R. Cope Jr., "A Model Municipal Department," 459–89 and 631–71, esp. 658. Also see comments by John S. Billings Jr. in Charles Rosenberg, "Making It in Urban Medicine," 173–76.

18. W. H. Guilfoy, "The Death Rate of the City of New York as Affected by the Cosmopolitan Character of Its Population," 132–35. Guilfoy, a physician, was the registrar of records for New York City.

19. Ibid., 134.

20. Antonio Stella, "The Effect of Urban Congestion on Italian Women and Children," 722–32. Stella surveyed streets in Harlem bounded by East 112th Street, East 113th Street, 1st Avenue, and 2nd Avenue, and blocks bounded by East Houston, Prince, Mott, and Elizabeth Streets, as well as 174 families from other sections of Manhattan.

21. Ibid. The overall death rate from diphtheria for children under five years for the city was 2.80/1000; the rate in the Italian section ranged from 3.00/1000 to 8.93/1000. The city's population was estimated at 2,564,432 when these data were collected.

22. Ibid., 724.

23. Recent studies by historians Richard Meckel and Deborah Dwork, and demographers Gretchen Condran, Ellen Kramarow, Samuel Preston, and Michael

Haines have all analyzed the differences in child mortality rates in New York City and other eastern cities at the turn of the century. Preston and Haines documented the ways in which child mortality varied according to the economic status of the family among the various immigrant groups and native-born American Blacks and whites. They confirmed that New York City presented one of the most deadly urban environments for children living in its tenement districts. It is very difficult to gain much information from this work on the specific mortality due to diphtheria, however, other than the fact that by 1900, diphtheria still accounted for over 5 percent of total deaths among children in the Death Registration Area. This attests to the fact that the use of antitoxin was not as uniform as Biggs and Park claimed. See Samuel Preston and Michael Haines, *Fatal Years*, 19. Also see Deborah Dwork, "Health Conditions of Immigrant Jews on the Lower East Side of New York, 1880–1914," 1–40. Using data from the 1910 census, Gretchen Condran and Ellen Kramarow confirm that the high Italian child mortality was associated with high fertility ("Child Mortality Among Jewish Immigrants to the United States," 250–53). It has still been difficult for scholars to explain the group differences in mortality rates in this period. For the influence of breast-feeding practices, see Richard Meckel, *Save the Babies*.

24. Stella, "Effect of Urban Congestion," 730.

25. Annie S. Daniel, "The Wreck of the Home," 624–29.

26. John Duffy, *History of Public Health in New York*, 265.

27. Cope, "Model Municipal Department," 651–53. Also see discussion in Duffy, *History of Public Health in New York*, 252–53.

28. See Duffy, chapters 11 and 12, for a more complete discussion of the shifting fortunes of the health department in the years 1900–1930.

29. The sale of antitoxin to a limited number of outside vendors resumed in 1904 for a time. In 1918 Park's lab also supplied antitoxin and other biologicals to the Federal Government to support the war effort; however, increased production also stretched his department's personnel to the limit. See Oliver, *The Man Who Lived for Tomorrow*, 214–43 and 364.

30. William Park, "A Critical Study of the Results of Serum Therapy in the Diseases of Man," 131.

31. In this work Ehrlich defined active immunity as the immunity acquired as a result of treatment or encounters with toxins, whereas passive immunity resulted from the injection of antitoxin obtained from another animal. See papers in *Collected Papers of Paul Ehrlich*, 3:86–106.

32. William Park, "Toxin-Antitoxin Immunization Against Diphtheria," 1584.

33. Ibid. The work Park referred to is E. Wernicke, "Ueber die Vererbung der künstlich erzeugte Diphtheri-Immunitat bei Meerschweinchen," Festschrift z. 100-Jahr. Stiftungsfeier des med.-Chir. Friedrich-Wilhelm-Inst., Berlin, 1895, 525.

34. William H. Park, "The Production of Diphtheria Antitoxin," *Proceedings of the New York Pathological Society*, 3 (May 1903): 139–40.

35. See Theobald Smith, "The Degree and Duration of Passive Immunity to Diphtheria Antitoxin Transmitted by Immunized Female Guinea Pigs to Their Immediate Offspring," 376, and "Active Immunity Produced by So-called Balanced or Neutral Mixtures of Diphtheria Toxin and Antitoxin," 241–56.

36. Smith, "Degree and Duration," 378.

37. William Park, "Production of Diphtheria Toxin," 139–40.

38. Emil Von Behring, "Über ein neues Diphtherieschutzmittel," *Deutsch. med. Wchnschr.* 39 (1913): 873. Behring received the first Nobel Prize in Physiology and Medicine in 1901 for his work on serum therapy, especially its application against diphtheria. See his discussion of the theoretical discussions on immunity in his acceptance speech in *The Nobel Lectures: Physiology and Medicine, 1901–1921* (Amsterdam: Elsevier Publishing Co., 1967), 3–18. Park argued in 1924 that though Behring is credited with the first use of toxin-antitoxin in humans, he neither discovered the method nor established the fact that those who had no antitoxin later developed it. See Park, Schroeder, and Zingher, "The Control of Diphtheria," 25 n.14.

39. Behring had not reported the definite proportions of toxin and antitoxin that were to be combined. See Park and Zinger, "Active Immunization with Diphtheria Toxin-Antitoxin," 3.

40. Park, Schroeder, and Zingher, "Control of Diphtheria," 25.

41. William Park, "The Possibility of Eliminating the Deleterious While Retaining the Antitoxic Effects of Antitoxic Sera," 308–88.

42. Antoni Gronowicz, *Béla Schick and the World of Children*, 77–82.

43. As Dowling notes, harmful or potential reactions following the injection of animal serums into humans had been observed as early as 1667. As discussed in Chapter 3, injections of antitoxin often caused rashes and swelling of joints. These symptoms usually were mild and passed quickly, though in some patients these reactions were more severe, causing serious discomfort. Von Pirquet and Schick made the first detailed study of these reactions, attributing them to an interaction between a foreign protein in the injected serum and an antibody in the patient's tissues. See Dowling, *Fighting Infection*, 38–39.

44. Schick and Pirquet named the phenomenon of hypersensitivity to foreign substances in animal serums *anaphylaxis*, meaning "against protection." See Clemens von Pirquet and Béla Schick, *Serum Sickness*.

45. The Schick test: A small thin syringe is filled with toxin and injected into the skin. Within 24 to 48 hours a reaction develops at the site of the injection. The skin becomes reddened. This positive result indicates that the person has no protective antibodies against diphtheria (no antitoxin). The negative result, no reaction, proves

the presence in the blood of antitoxin in an amount sufficient to protect the person from diphtheria. This description of the Schick test is taken from the translation of his paper, "The Skin Reaction with Diphtheria Toxin on Human Beings as a Test Preceding the Prophylactic Injection of Diphtheria Serum," in Gronowicz, *Béla Schick*, 78–82 (first published in *Muenchener Medizinische Wochenschrift*, 60 [1913]: 2608).

46. Gronowicz, *Béla Schick*, 81.

47. Paul de Kruif, *The Microbe Hunters*, 206.

48. William Park and Abraham Zingher, "Practical Reactions Obtained from the Schick Reaction," 151–58.

49. Abraham Zingher (1885–1927) was placed in charge of the active immunization campaign under Park. Born in Rumania, he immigrated to the United States as a boy. He completed his medical degree at Cornell. During World War I he served as a captain in the Medical Corps and became supervisor of bacteriological laboratories in France. In the New York City Health Department he served as assistant director of the research laboratory under Park and also as professor of hygiene at New York University Medical School, professor of clinical medicine at Polyclinic Hospital, and pediatrist at the Post-Graduate Hospital. See obituary in the *New York Times*, 6 June 1927, 1.

50. William Park, Abraham Zingher, and Harry M. Serota, "Active Immunization in Diphtheria and Treatment by Toxin-Antitoxin," 859; and "Active Immunization with Diphtheria Toxin-Antitoxin — Second Paper," 2217.

51. "Active Immunization — Second Paper," 2219.

52. William Park, "Experience With the Diphtheria Toxin Skin Reaction," 417.

53. Park and Zingher, "Practical Application Obtained from the Schick Reaction," 151–52. Park's description of the Schick reaction appears a bit vague in comparison to others. We might contrast his description with that of a Sheldon Dudley, a British physician: "In about twenty-four hours, or earlier after the injection of a Schick dose of toxin, a red, almost scarlet patch about an inch in diameter appears, which is generally elliptical rather than circular; the major axis of the ellipse being in the line of the limb. In four or five days the reaction after first becoming more intense, begins to fade and desquamate, and becomes covered with bran-like flakes of cracked superficial epithelium which gives to the reaction most characteristic appearance at the end of the week" ("Critical Review: Schick's Test and Its Applications," 323).

54. See questions put to Park at the Practitioners' Society of New York, April 2, 1915, printed in the *Medical Record*, 18 September 1915, 505–06.

55. William Park, "Diphtheria," *Boston Medical and Surgical Journal*, 722.

56. In the control test, heated toxin or fully neutralized toxin was injected into the

arm opposite from the test arm in order to differentiate between the true and the pseudoreaction. The assumption was that the pseudoreaction was due to the autolyzed protein of the diphtheria bacillus and not to the soluble toxin that gave rise to the true reaction. This protein represented the body substance of the diphtheria bacilli that had disintegrated and become dissolved in the broth culture used for making the toxin. By heating the test toxin, the soluble toxic fraction was destroyed, leaving the autolyzed protein. It could be determined whether the person was reacting to the protein, since the pseudoreaction was found to disappear after 48 hours, while the true reaction persisted. See A. Zingher, "Interpretation of the Schick Reaction in Recruits for the National Army," 228.

57. Abraham Zingher, "Methods of Using Diphtheria Toxin in the Schick Test and of Controlling the Reaction," 272.

58. William Park, Abraham Zingher, and Harry M. Serota, "The Schick Reaction and Its Practical Applications," 481.

59. Zingher, "Methods of Using Diphtheria Toxin," 275. It was widely believed that diphtheria carriers always had a negative Shick reaction. By 1929 researchers recognized that this was not always the case, and they remained puzzled as to the role of the virulence of the diphtheria bacilli. See Dudley, "Critical Review: Schick's Test," 350.

60. Among the institutions where the trials of the Schick test were conducted were the Home for Colored Orphans, St. John's Home in Brooklyn, the Hebrew Orphan Asylum, New York Foundling Home, New York Catholic Protectory, Randall's Island Hospital, and Children's Stations run by the Bureau of Child Hygiene. See M. C. Schroeder, "The Duration of the Immunity Conferred by the Use of Diphtheria Toxin-Antitoxin," 369–72.

61. For example, Schroeder reported on tests on 4,000 inmates of the State Institution for the Insane, "in order to obtain results on a group of people likely to remain located indefinitely." See M. C. Schroeder, "The Duration of Active Immunity after Injections of Diphtheria Toxin-Antitoxin," 437.

62. Mr. George Bedinger, comments in the *Medical Record*, 3 September 1921, 436.

63. Abraham Zingher, "Diphtheria Preventive Work in the Public Schools of New York City," *Medical Record*, 3 September 1921, 438.

64. Abraham Zingher, "Diphtheria Prevention Work in the Public Schools of New York City," *Journal of the American Medical Association*, 77, no. 11 (1921): 836.

65. Park, Schroeder, and Zingher, "Control of Diphtheria," 26.

66. Dr. Dever S. Byard, "The Schick Test and Active Immunization with Toxin-Antitoxin in Private Practice," 437.

67. "Symposium on Modern Methods for the Prevention of Diphtheria," *Medical Record*, 3 September 1921, 439.

68. The report noted that the previous year the department had encountered some problems for overstating the value of a typhoid vaccine (Annual Report of the Board of Health for 1922, 83).

69. Park, Comments at Practitioner's Society of New York.

70. Zingher, "Diphtheria Prevention Work in the Public Schools," 835–43.

71. In 1917, Dr. Louis T. Wright of Freedmen's Hospital in Washington, D.C., questioned Zingher's interpretation of the Schick test results in Black populations. Wright noted that the interpretation of the results rested on identifying reddened areas and discolored areas of the skin after the injection. Wright noted that these descriptions applied only to whites and that the degree of pigmentation in Blacks could affect the reading of the results. In addition he noted that Zingher had failed to discuss the degree of natural immunity that Blacks possessed to diphtheria based on the test. In his test at Freedmen's, he found that adult Blacks had the same immunity to diphtheria as the whites in his study. See Louis T. Wright, "The Schick Test, with Especial Reference to the Negro," 265–68.

72. Zingher, "Diphtheria Prevention Work in the Public Schools," 839.

73. Ibid., 838.

74. This point is made by Haven Emerson, 317. The following discussion draws heavily from his 1931 paper, "Significant Differences in Racial Susceptibility to Measles, Diphtheria and Scarlet Fever," 317–50.

75. Ibid., 339. The mortality rate for the Irish was 164.3/100,000 and for Blacks 149.5/100,000. Morbidity per 100,000 was 513.6 for the Irish and 457.8 for the Black population.

76. Ibid., 346.

77. Researchers at Johns Hopkins were surprised at Zingher's findings of a high number of Blacks in New York City with no immunity to diphtheria. The prevailing view among health experts in the South was that Blacks were immune to diphtheria. Researchers at Hopkins continued to pursue this question into the 1930s. See J. B. Black, "A Comparative Study of Susceptibility to Diphtheria in the White and Negro Races," 734–48; and James Doull, "Factors Influencing Selective Distribution in Diphtheria," 371–403.

78. Abraham Zingher, "Active Immunization of Infants Against Diphtheria," 83–102.

79. Ibid., 102.

80. The Boston accident generated a great deal of publicity and an investigation by the Massachusetts Department of Public Health. Park was a member of the group

that investigated the accident, which concluded that no one was at fault. See "Frozen toxin-antitoxin reactions — Letter to Editor," *Journal of the American Medical Association*, 82 (16 February 1924): 567–68; and William Park, "Some Important Facts Concerning Active Immunization Against Diphtheria," 709–17.

81. In a survey of all aspects of the Schick test, Dudley noted the critical need for using standardized test toxin ("Critical Review: Schick Test," 331). He also noted that other workers reported more difficulty in reading the reactions than did Park and his colleagues. He noted the need for experienced workers to read the test results (332 and 334).

82. "Gas Kills Scientist in Quest for Serum," *New York Times*, 6 June 1927, 1, 3.

83. "Immunization," Editorial, *New York Times*, 7 June 1927, 28:3.

84. Much more archival work is needed using records of the various institutions where Park and Zingher conducted the active immunization trials to find out how they were conducted and the problems encountered. Oliver, Park's biographer, is curiously silent on the details of this work. See also references to criticism of Park's methods in the 1930s in Susan Lederer, *Subjected to Science: Human Experimentation in America Before the Second World War*, 108–9 and 137.

85. Lederer, *Subjected to Science*, 137.

86. See discussion of Zingher's and Park's work on polio in Naomi Rogers, *Dirt and Disease: Polio Before FDR*, 99–102.

87. Harris, "Symposium on Modern Methods," 440.

88. Comments of Edward L. Bauer following Zingher, "Diphtheria Prevention Work," 842.

89. See the story recounted in Oliver of one of Park's associates who was involved in the early study of the Schick test. A Dr. Blum performed the test on a child from the tenement district in the absence of the mother, who later made a protest to the health department. The mother was appeased, and later Dr. Blum was complimented on his "energy and courage" (*The Man Who Lived for Tomorrow*, 324).

90. The two essays where this is most evident have been cited. See his 1922 article, "The Control of Diphtheria," and his "The History of Diphtheria in New York City" (1931).

Chapter 7: *The Diphtheria Prevention Commission*

1. For a discussion of other public health campaigns in the Progressive Era, see Martin S. Pernick, "Thomas Edison's TB Films: Mass Media and Health Propaganda," 21–27.

2. George Bedinger in the discussion in "Forum on Modern Methods for the Prevention of Diphtheria," *Medical Record*, 3 September 1921, 440.

3. C.-E. A. Winslow and S. Zimand, *Health Under the El: The Story of the Bellevue-Yorkville Health Demonstration Project*, 33.

4. Ibid., 188.

5. George Rosen, "The First Neighborhood Health Center Movement," 188.

6. Ibid., 189.

7. Winslow and Zimand, *Health Under the El*, 35.

8. Ibid., 14.

9. Ibid., 27.

10. Godias J. Drolet and E. Clark, *The Bellevue-Yorkville Health Demonstration, New York City*, 19, 20. Columbia University Archives.

11. Winslow and Zimand, *Health Under the El*, 26.

12. S. Josephine Baker, *Fighting For Life*, 160.

13. Ibid., 152.

14. All comments are from "Report on Bellevue-Yorkville Demonstration, June 1930," Milbank Memorial Fund Papers, Sterling Library, Yale University, Box 31, Series III, Group No. 845.

15. For a discussion of reformers' complex response to the issue of parental authority, see Linda Gordon, "Family Violence, Feminism, and Social Control," 146–47.

16. See Michael M. Davis, *Immigrant Health and the Community*.

17. *Annual Report Board of Health*, 1924, 39.

18. See Rima Apple, *Mothers and Medicine: A Social History of Infant Feeding, 1890–1950*, chap. 6; Molly Ladd-Taylor, "Hull House Goes to Washington: Women and the Children's Bureau," 110–26; and Richard Meckel, *Save the Babies*.

19. Preston and Haines, *Fatal Years*, 35.

20. *Monthly Bulletin*, Department of Health, 15, no. 11 (13 March 1926): 42.

21. See *Weekly Bulletin*, Department of Health, 16, no. 8 (19 February 1927): 29.

22. *Monthly Bulletin*, Department of Health, 16, no. 21 (31 May 1927): 82–83.

23. Editorial, *New York Medical Week*, 7 (22 December 1928): 6.

24. Ibid.

25. "Wynne Maps Drive to End Diphtheria," *New York Times*, 12 January 1929, 19.

26. The members of the Diphtheria Prevention Commission included some of New York's most prominent citizens. Among them were Edward F. Brown, director; Mrs. August Belmont, Cornelius N. Bliss, Nicholas Murray Butler, Ph.D., Haley Fiske, Lee K. Frankel, Ph.D., Mrs. Charles Dana Gibson, Cardinal Patrick Hayes, Frederic A. Juilliard, John A. Kingsbury, Herbert Lehman, Albert G. Milbank, Jeremiah Milbank, and Ralph Pulitzer. See *Annual Report Board of Health*, 1928, 13, for complete list.

27. Ibid. The Technical Consultation Board of the Diphtheria Prevention Commission consisted of Béla Schick, William Park, and Edward L. Benjamin, the Bronx Medical Society; Walter F. Watton, Kings County Medical Society; Walter L. Carr, New York County Medical Society; George Walrath, Richmond County Medical Society; Philip Stimson, president of the School Physicians' Association; William H. Guilfoy, statistical consultant; and Alec Thomson, secretary of the Board of Health.

28. Ibid.

29. The executive director of the commission, Edward F. Brown, raised $70,000 each year of the three years of the campaign through the Milbank Fund, and an equal amount was contributed from the health department's fund. In total about $140,000 per year was dedicated to the campaign. See New York Times, 8 November 1931.

30. Editorial, "No Excuse for Diphtheria," New York Times, 14 January 1929, 22.

31. "Diphtheria Drive Gains," New York Times, 22 January 1929, 31.

32. Susan Sontag, AIDS and Its Metaphors, 10–15.

33. "Diphtheria Deaths Cut By Campaign," New York Times, 10 March 1929, 1.

34. "Doctors Envision End of Diphtheria," New York Times, 16 March 1929.

35. Nathan B. Van Etten, "The Six Hundred and Fiftieth Man in Diphtheria Prevention," New York State Journal of Medicine, 15 April 1929, 452.

36. Quoted in Zingher, "Diphtheria Prevention Work in the Public Schools," 841. For a discussion of some aspects of the antivivisectionist movement and the use of children in medical experimentation see Susan Lederer, Subjected to Science: Human Experimentation in America Before the Second World War (Baltimore: Johns Hopkins University Press, 1995), chap. 5.

37. "Widens Diphtheria Drive: Dr. Wynne Says City Physicians Will Give Aid at All Hours," New York Times, 28 March 1992.

38. See John Duffy, "The American Medical Profession and Public Health," 1–15.

39. "Wynne Acts to Widen Diphtheria Drive," New York Times, 27 May 1929, 20. There is some indication that physicians' resistance to the diphtheria campaign had to do with the cost of immunization versus the cost of treating cases. Some pointed out, however, that with respect to diphtheria, income from immunizations was comparable to if not greater than that from treatment of diphtheria cases. See "The Doctor's Income and Preventive Medicine," Editorial, American Journal of Public Health, 1 (January 1939): 50–51.

40. Edward T. Devine, "New York City's Diphtheria Campaign," 26.

41. See Lillian Wald, The House on Henry Street, chap. 2.

42. Editorial, "Public Health and the Family Doctor," New York Medical Week, 8, no. 5, (2 February 1929): 7.

43. "New York's Diphtheria Prevention Campaign," New England Journal of Medicine (25 April 1929): 898.

44. "Doctors Envision End of Diphtheria," *New York Times*, 16 March 1929, 8.

45. Diane Hamilton, "The Cost of Caring: The Metropolitan Life Insurance Company Visiting Nurse Service, 1909–1953," 419.

46. Lee K. Frankel, "No Diphtheria in New York State in 1930," Speech presented at the Annual Meeting of the State Committee on Public Health of the State Charities Aid Association, May 5, 1925. Metropolitan Life Insurance Company Archives, New York (hereafter MLIC Archives), 6–7.

47. See Dorothy Nelkin, "AIDS and the News Media," 293–307.

48. Shirley W. Wynne, "Banishing Diphtheria as a Scourge of Childhood," *Annual Report Board of Health*, 1928, 21.

49. Michael Schudson attributes this emphasis on public relations by the press to the success of campaigns during World War I for war bonds. Following that, organizations like the Red Cross, the Salvation Army, and the YMCA developed publicity campaigns based on the wartime model. See Schudson, *Discovering the News*, chap. 4.

50. "Thirty Stations in Parks to Fight Diphtheria," *New York Times*, 1 May 1929.

51. "Clinics on Wheels Ready to Tour City," *New York Times*, 13 July 1929.

52. Wynne, "Banishing Diphtheria," 26.

53. "Cardinal Hayes Aids Diphtheria Campaign," *New York Times*, 8 April 1929, 52.

54. *Weekly Bulletin*, Department of Health, no. 27, 6 July 1929.

55. Iago Galdston, "Diphtheria," Radio Talk, MLIC Archives, New York (not dated but refers to campaign in text). Dr. Galdston was head of the Medical Information Bureau of the New York Academy of Medicine. Its purpose, according to Philip van Ingen, was "to control information in the lay press within reasonable and accurate bounds. It tried to advise on medical views and claims received by the daily press, the value and soundness of which the press had no way of judging." It established a sort of supervision of radio addresses on health matters and gave an annual dinner for the press. See Van Ingen, *The New York Academy of Medicine: Its First Hundred Years* 434–435.

56. William Park's version of this story was also given as a radio talk and was published by him in W. H. Park and Anna W. Williams, *Who's Who Among the Microbes*, 164–78.

57. Wynne, "Banishing Diphtheria," 27.

58. The editorial emphasized that the British were rightly taking a more cautious approach toward the widespread use of toxin-antitoxin. Editorial, "Toxin-Antitoxin in Diphtheria Immunization," *British Medical Journal*, 6 July 1929, 22–23.

59. George F. and Gladys H. Dick, "Immunization Against Diphtheria: Comparative Value of Toxoid and Toxin-Antitoxin Mixtures," 1901–3.

60. William Park, "The History of Diphtheria in New York City," 1439–45.

61. New York's immunization program under Park's direction had few technical problems, but between 1919 and 1930 there were several incidents where toxin-antitoxin mixtures became contaminated or were stored at improper temperatures, resulting in cases of severe reaction and some deaths. See Parish, *A History of Immunization*, 151–53.

62. William Park, "Some Important Facts Concerning Active Immunization Against Diphtheria," 2:17.

63. Elwin Nash and J. Graham Forbes, "Diphtheria Immunization: Its Possibilities and Difficulties," *Public Health*, 46 (October–September 1933): 253.

64. Edward F. Brown "New York City's Diphtheria Campaign," 22–23.

65. "Brown Ends Drive to Halt Diphtheria," *New York Times*, 8 November 1931.

66. *Annual Report Board of Health*, 1930–1931, 186.

67. Devine, "New York City's Diphtheria Campaign," 27. The Harmon Award was given by the Harmon Foundation, a New York–based foundation, to recognize achievements in social and philanthropic work. See *Bulletin of the Russell Sage Foundation* (August 1926): 19–20.

68. "A Preventable Disease," Editorial, *New York Times*, 16 May 1933, 16.

69. Michael B. Katz, *In the Shadow of the Poorhouse: A Social History of Welfare in America*, 142.

70. C.-E. A. Winslow, *The Evolution and Significance of the Modern Public Health Campaign*, 57–58.

71. Richard Meckel makes a similar point with respect to the negative and positive benefits of maternal and child health campaigns that were contemporaneous with the diphtheria prevention campaign. See Meckel, *Save the Babies*, 220–21.

Epilogue

1. H. M. Bisgard et al., "Virtual Elimination of Respiratory Diphtheria in the United States," 280.

2. Bernard Dixon, "Infectious Diseases — A 'Dead' Specialty Revives," 779.

3. International data from World Health Organization (WHO) *HIV/AIDS Sexually Transmitted Diseases Newsletter* (December 1997 to March 1998), 2. Through June 1997 a cumulative total of 612,078 persons with AIDS had been reported to the Centers for Disease Control (CDC). *HIV/AIDS Surveillance Report*, vol. 9, no. 1, 1997, p. 3.

4. Historians have responded to these questions from a number of perspectives. Two books that exemplify the range of these responses are Elizabeth Fee and Daniel

M. Fox, eds., *AIDS: The Burdens of History*; and Fee and Fox, *AIDS: The Making of a Chronic Disease*.

5. See Allan Brandt, "AIDS: From Social History to Social Policy," in *AIDS: The Burdens of History*, 163.

6. For a somewhat different analysis of the lessons to be drawn from the history of diphtheria, see Lawrence C. Kleinman, "To End an Epidemic: Lessons from the History of Diphtheria," 773–77.

7. See Hammonds, "Race, Sex, AIDS: The Construction of 'Other,'" 28–36, and "Missing Persons: African-American Women, AIDS, and the History of Disease," 7–24. See also Nicholas Freudenberg, "AIDS Prevention in the United States," 589–99.

8. Freudenberg, "AIDS Prevention in the United States," 591.

9. Sheryl Gay Solberg, "AIDS Drugs Give Little Hope to Thousands Unable to Pay," *New York Times*, 14 October 1997, 1, A22.

10. For current discussions of these issues, see Paul Greenough, "Global Immunization and Culture: Compliance and Resistance in Large-Scale Public Health Campaigns," 605–7.

BIBLIOGRAPHY

One of the most difficult problems facing scholars doing research on the New York City Health Department before the turn of the century is the absence of the papers of two of its most prominent figures, Hermann Biggs and William H. Park. The biographies of these men, C.-E. A. Winslow's *The Life of Hermann M. Biggs* (1929) and Wade W. Oliver's *The Man Who Lived for Tomorrow* (1941), are excellent sources but must be used carefully because in both cases the biographers' views are infused into the narratives. I have tried in as many instances as possible to supplement these biographers' reconstructions of events with primary documents and newspaper articles.

Manuscript Sources

Columbia University Archives, New York, N.Y.: Charity Organization Society Papers.

John Hay Library, Brown University Archives. Providence, R.I.: Charles V. Chapin Papers.

Metropolitan Life Insurance Company Archives, New York, N.Y.: Haley Fiske Papers; Lee K. Frankel Papers; Health and Safety Files.

New York Academy of Medicine Archives, New York, N.Y.: Minutes of the Medical Board of the Willard Parker Hospital; Minutes of the Medical Society of the County of New York, 1894–1898.

Rhode Island Historical Society, Providence, R.I.: Charles V. Chapin Papers.

Schlesinger Library, Radcliffe College, Cambridge, Mass.: Mary Putnam Jacobi Papers; Anna Wessels Williams Papers.

Yale University Archives, New Haven, Conn.: Milbank Memorial Fund Papers; T. Mitchell Prudden Papers; C.-E. A. Winslow Papers.

Government and Other Official Publications

Billings, John S., *Vital Statistics of New York and Brooklyn: Covering a Period of Six Years Ending May 31, 1890*. Washington, D.C.: Government Printing Office, 1894.

City of New York, Department of Health, *Monthly Bulletin*, 1920–1927.

City of New York, Department of Health, *Weekly Bulletin*, 1926–1929.

Claghorn, Kate H. "The Foreign Immigrant in New York City." Report of the Industrial Commission 15. Washington, D.C.: Government Printing Office, 1901.

Collected Studies from the Research Laboratory, Department of Health, City of New York, 1905.

Connecticut State Board of Health, *Annual Report*, 1895.

Health Department, City of New York, Annual Reports of the Board of Health, 1874–1931.

Minutes of the New York City Board of Health, 1893–1896.

New York City Department of Health, Monograph Series, 1912–1920.

New York City Health Department Bacteriological Laboratory, *Scientific Bulletin*, nos. 1–2, 1894–1895.

Report of the New York State Tenement House Committee as authorized by Chapter 479 of the Laws of 1894. Transmitted to the Legislature January 17, 1894. (Tenement House Report)

Newspapers

New York Herald, 1894–1900.

New York Times, 1885–1930.

Unpublished Material

Atkins, Gordon. "Health Housing and Poverty in New York City, 1865–1898." Ph.D. diss. Columbia University. Ann Arbor Michigan: Gordon Atkins, 1947.

Blancher, David. "Workshops of the Bacteriological Revolution: A History of the Laboratories of the New York City Department of Health, 1892–1912." Ph.D. diss., City University of New York, 1979.

Gossel, Patricia Peck. "The Emergence of American Bacteriology, 1875–1900." Ph.D. diss., The Johns Hopkins University, 1989.

Harkness, Jon M. "Rabies and Research: A History of the New York Pasteur Institute." Paper presented at the American Association for the History of Medicine Annual Meeting, Birmingham, Alabama, 1989.

Howson, Jean E. "Sure Cure for Diphtheria: Medicine and New York City Newspapers in 1894." MS, New York University, 1986.

Skolnik, Richard. "The Crystallization of Reform in New York City, 1890–1917." Ph.D. diss., Yale University, 1964.

Primary Published Sources

Abt, Isaac, ed. *Pediatrics*. Philadelphia: W. B. Saunders, 1923–1926.

Acer, Louis C. "The First Four Months of Diphtheria Cultures in Brooklyn," *Brooklyn Medical Journal* 9 (1895): 93.

Andrewes, Frederick, et al., ed. *Diphtheria: Its Bacteriology, Pathology and Immunology*. London: His Majesty's Stationery Office, 1923.

Baker, S. Josephine. *Fighting For Life*. Huntington, N.Y.: Robert E. Krieger Publishing Co., 1980. Reprint of 1939 edition (New York: Macmillan).

Behring, E. "Über ein neues Diphtherieschutzmittel." *Deutsche med. Wchnsch.* 39 (1913): 873.

———."Untersuchungen über das Zustandekommen der Diphtherie-Immunität bei Thieren." *Deutsche med. Wchnsch.* 16 (1890): 1145–48.

Behring, E., and S. Kitasato. "Über das Zustandekommen der Diphtherie-Immunität und der Tetanus-Immunität bei Thieren." *Deutsche med. Wchnsch.* 16 (1890): 113–14.

Biggs, Hermann. "The Health of the City of New York." *Transactions of the New York Academy of Medicine*, 2d ser., 12 (1895): 420–48.

———. "History of the Recent Outbreak of Epidemic Cholera in New York." *American Journal of Medical Sciences*, n.s., 105 (January 1893): 63–72.

———. "The New Treatment of Diphtheria." *McClure's Magazine* 4 (1895): 360.

———. "The Organization, Equipment, and Method of Work of the Division of Pathology, Bacteriology and Disinfection of the New York City Health Department." *Transactions of the New York Academy of Medicine*, 2d ser., 10 (1893): 303–20.

———. "Preventive Medicine in the City of New York." *British Medical Journal* 2 (11 September 1897): 629–38.

———. "Sanitary Science, the Medical Profession, and the Public." *Medical News* 72 (8 January 1898): 1635–40.

———. "The Serum Treatment and Its Results." *Medical News* 75 (22 July 1899): 97.

———. "Some Experiences in the Production and Use of Diphtheria Antitoxin." *Medical Record* 47 (20 April 1895): 481–84.

———. "Some Investigations as to the Virulence of the Diphtheria Bacilli Occasionally Found in the Throat-Secretions in Cases Presenting the Clinical Features of Simple Acute Angina." *Transactions of the Association of American Physicians* 11 (1896): 249.

———. "The Use of Antitoxic Serum for the Prevention of Diphtheria." *Medical News* 67 (1895): 589.

Billings, John S. "Municipal Sanitation in New York City." *The Forum* 16 (1893): 347.

Billings, John S., Jr. "Ten Years Experience with Diphtheria Antitoxine." *New York Medical Journal* 82 (23 December 1905): 1310–12.

Billington, C. E. *Diphtheria: Its Nature and Treatment*. New York: William Wood and Co., 1889.

Black, J. B. "A Comparative Study of Susceptibility to Diphtheria in the White and Negro Races." *American Journal of Hygiene* 19 (1934): 734–48.

Byard, Dr. Dever S. "The Schick Test and Active Immunization with Toxin-Antitoxin in Private Practice." *Medical Record*, 3 September 1921, 437.

Caillé, Augustus. "A Method of Prophylaxis in Diphtheria." *Medical Record* 33 (February 1888): 183–86.

Carey, Bernard. "Diphtheria, the Uncontrolled." *Boston Medical and Surgical Journal* 181 (1919): 92–95.

———. "Diphtheria Control." *Journal of the American Medical Association* 77 (27 August 1921): 668–72.

Chapin, Charles V. "The Restriction of Contagious Diseases in Cities." *American Medicine*, 10 (1905): 992–95.

———. *The Sources and Modes of Infection*. New York: John Wiley and Sons, 1910.

Chappell, Walter. "Vexed Questions in the Bacteriology of Diphtheria." *Medical Record*, 14 April 1894, 457.

Cope, Francis R., Jr. "A Model Municipal Department: How the Low Administration Has Been Caring for the Health of New York City." *American Journal of Sociology* 9 (1903): 459–89, 631–71.

Craig, Charles. "Unrecognized Infections in Production of Carriers of Pathogenic

Microorganisms." *Journal of the American Medical Association* 77 (10 September 1921): 827–33.

Crum, Frederick S. "A Statistical Study of Diphtheria." *American Journal of Public Health* 7 (1916): 445–77.

Daniel, Annie S. "The Wreck of the Home." *Charities* 14 (1905): 628–29.

Denny, Francis P. "Prize-winning Essay." *Journal of the Massachusetts Association of Boards of Health* 10 (July 1900): 55–72.

Devine, Edward T. "New York City's Diphtheria Campaign." *Milbank Memorial Fund Quarterly Bulletin* 9 (January 1931): 20–27.

Dick, George, and Gladys H. Dick. "Immunization Against Diphtheria." *Journal of the American Medical Association* 92 (1929): 1901–03.

Discussion on Diphtheria Antitoxin. Association of American Physicians, 10th Annual Meeting, Washington, D.C., May 30–31, 1895. *Boston Medical and Surgical Journal* 132 (27 June 1895): 645–49.

Discussion on the Use of Diphtheria Antitoxin. *Transactions of the New York Academy of Medicine*, 4 April 1895, 204–215.

Doull, James. "Factors Influencing Selective Distribution in Diphtheria." *Journal of Preventive Medicine* 4, no. 5 (1930): 371–403.

Doull, James A., and H. Lara. "The Epidemiological Importance of Diphtheria Carriers." *American Journal of Hygiene* 5, no. 4 (1925): 508–29.

Drolet, Godias, and E. Clark. *The Bellevue-Yorkville Health Demonstration Project, New York City: A Reference Handbook on Births, Deaths and the Prevalence of Disease.* 1931. [s.n.]

Dunham, Edward K. "The Bacteriological Examination of the Recent Cases of Epidemic Cholera in the City of New York." *American Journal of Medical Sciences,* n.s. 105 (January 1893): 72–80.

Ehrlich, Paul. *Collected Papers of Paul Ehrlich,* ed. F. Himmelweit, M. Marquardt, and Henry Dale. London: Pergamon Press, 1960.

Emerson, Haven. "Significant Differences in Racial Susceptibility to Measles, Diphtheria and Scarlet Fever: A Statistical Study of Case Incidence and Deaths Among Tenement Dwellers in New York City, 1921–1925, Inclusive," *Journal of Preventive Medicine* 5 (September 1931): 317–50.

Frobisher, M., E. Parsons, and E. Updike. "The Correlation of Laboratory and Clinical Evidence of Virulence of C. Diphtheriae." *American Journal of Public Health* 37 (1947): 543–48.

Frost, Wade H. "Infection, Immunity and Disease in the Epidemiology of Diphtheria with Special Reference to Some Studies in Baltimore," *Journal of Preventive Medicine* 2 (July 1928): 325–43.

Frost, Wade H., M. Frobisher Jr., V. A. Van Volkenburgh, and M. L. Levin. "Diphtheria in Baltimore: A Comparative Study of Morbidity, Carrier Prevalence, and Antitoxic Immunity in 1921–1924 and 1933–1936." *American Journal of Hygiene* 24 (November 1936): 568–86.

Gibier, Paul. "Further Notes Upon the Treatment of Diphtheria with the Antitoxin Made at the New York Pasteur Institute With Report of Cases." *New York Therapeutic Review* 3 (June 1895): 29–38.

———. "The Treatment of Diphtheria with the Antitoxine Made at the New York Pasteur Institute with Report of Cases." *New York Therapeutic Review* 3 (March 1895): 1–00.

Guilfoy, W. H., "The Death Rate of the City of New York as Affected by the Cosmopolitan Character of Its Population." *Medical Record* 73 (25 January 1908): 132–35. (Also published in *Quarterly Publications of the American Statistical Association* 10 [1906–07]: 515–22.)

Harris, Louis I. "Symposium on Modern Methods for the Prevention of Diphtheria," *Medical Record*, 3 September 1921, 439.

Hill, Hibbert W. "The Official Definition of Diphtheria." *Public Health Reports* 25 (1899): 243–48.

Hirsch, August. *Handbook of Geographical and Historical Pathology*. 2d. ed. Translated by Charles Creighton. London: New Sydenham Society, 1886.

Hueppe, Ferdinand. *The Methods of Bacteriological Investigation*. Translated by Hermann M. Biggs. New York: D. Appleton and Co., 1896.

Jacobi, Abraham. *Collected Essays, Addresses, Scientific Papers and Miscellaneous Writings of Abraham Jacobi*. Edited by William Robinson. 8 vols. New York: Critic and Guide, 1909.

———. "Diphtheria." In *A System of Practical Medicine by American Authors*, ed. William Pepper, 1:696–97. Philadelphia: Lea Brothers Co., 1885.

———. "Diphtheria: Its Symptomatology and Treatment." In A. Jacobi, *Contributions to Pediatrics*, ed. William J. Robinson, 1:121–212. New York: Critic and Guide, 1909.

———. "Diphtheria Spread by Adults." *New York Medical Journal* (27 September 1884): 344–47.

———. "Follicular Amygdalitis." *Medical Record*, 27 November 1886, 593–600.

———. "A History of Pediatrics in New York." *Archive of Pediatrics*, 34 (1917): 5–6.

———. "Inaugural Address." *Transactions of the New York Academy of Medicine* (1885), 5. In *Collected Essays, Addresses and Scientific Papers and Miscellaneous Writings of Abraham Jacobi*, ed. William Robinson, 7:157–78. New York: Critic and Guide, 1909.

——. *On the Treatment of Diphtheria in America.* Berlin: L. Schumacher, 1891.

——. *A Treatise on Diphtheria.* New York: William Wood, 1880.

Loeffler, F., "Der gegenwartige Stand der Frage nach der Entstehung der Diphtherie." *Deutsche Medizinische Wochenschrift,* Nos. 5 and 6, 1890.

Moss, W. L. "Diphtheria Bacillus Carriers." Transactions of the 15th International Congress on Hygiene and Demography. Washington, D.C.: Government Printing Office, 1913, 4:165.

Nash, Elwin, and J. Graham Forbes. "Diphtheria Immunisation: Its Possibilities and Difficulties." *Public Health* 46 (1933): 245–71.

Nichols, Henry J. *Carriers in Infectious Diseases.* Baltimore: Williams and Wilkins, 1922.

Nuttall, G. H. F., and G. S. Graham-Smith, eds. *The Bacteriology of Diphtheria.* Cambridge: Cambridge University Press, 1908. Reissue with supplementary bibliography, 1913.

O'Dwyer, Joseph. *Intubation in Croup and Other Acute and Chronic Forms of Stenosis of the Larynx.* New York: William Wood, 1889.

Oliver, Wade W. *The Man Who Lived for Tomorrow.* New York: E. P. Dutton, 1941.

Park, William [H.]. "Active Immunity Produced by So-called Balanced or Neutral Mixtures of Diphtheria Toxin and Antitoxin." *Journal of Experimental Medicine* 11 (1909): 241–56.

——. "The Clinical Use of Diphtheria Antitoxin." *Boston Medical and Surgical Journal* 133 (12 September 1895): 256.

——. "Contribution of Bacteriology to Therapeutics." *Pediatrics* 4 (1 and 15 November 1897): 385–98, 433–49.

——. "A Critical Study of the Results of Serum Therapy in the Diseases of Man." In *The Harvey Lectures,* 1905–06, 131. Philadelphia: J. B. Lippincott Co., 1906.

——. "Diphtheria." *Boston Medical and Surgical Journal* 175 (16 November 1916): 721–23.

——. "Diphtheria." In *Public Health and Hygiene,* ed. William H. Park, 100–115. Philadelphia: Lea & Febiger, 1920.

——. "Diphtheria and Allied Pseudo-membranous Inflammations: A Clinical and Bacteriological Study." *Medical Record* 42 (30 July 1892): 113–25.

——. "Diphtheria and Allied Pseudo-membranous Inflammations—Part 2." *Medical Record* 42 (6 August 1892): 141–47.

——. "Diphtheria Immunity—Natural, Active and Passive: Its Determination by the Schick Test." *American Journal of Public Health* 6 (1916): 431–45.

——. "Diphtheria and Other Pseudo-membranous Inflammations: A Clinical and

Bacteriological Study—Second Paper." *Medical Record* 43 (11 February 1893): 161–67.

———. "Experience With the Diphtheria Toxin Skin Reaction." *Medical Record,* 6 March 1915, 417.

———. "Health Laboratory Organization," *Journal of the American Medical Association* 67 (30 December 1916): 2013–15.

———. "The History of Diphtheria in New York City." *American Journal of Diseases of Children* 42 (December 1931): 1431–45.

———. "The Possibility of Eliminating the Deleterious While Retaining the Antitoxic Effects of Antitoxic Sera." *Transcriptions of the American Association of Physicians* 15 (1900): 308–88.

———. "The Preparation of the Diphtheria Antitoxin and Some of the Practical Lessons Learned from the Animal Experiments Performed in Testing Its Value." *Medical Record* 47 (20 April 1895):485–86.

———. "The Production of Diphtheria Antitoxin." *Proceedings of the New York Pathological Society* 3 (May 1903): 139–40.

———. "The Relation of the Toxicity of Diphtheria Toxin to Its Neutralizing Value upon Antitoxin at Different Stages in the Growth of Culture." *Journal of Experimental Medicine* 3 (1898): 513–32.

———. "Some Factors Which Lead to the Increase and Decline of Communicable Diseases Among Men and Animals." *Health News,* New York State Department of Health, old ser., 38 (1923): 193.

———. "Some Important Facts Concerning Active Immunization Against Diphtheria." *American Journal of Diseases of Children* 32 (1926): 709–717.

———. "Some Recent Studies on the Communicability and Treatment of Diphtheria and Pseudo-diphtheria," *Transactions of the New York State Medical Association,* 11 (1894): 340–63.

———. "Toxin-Antitoxin Immunization Against Diphtheria." *Journal of the American Medical Association* 79 (4 November 1922): 1584–91.

Park, William H., and Alfred Beebe. "Report on Bacteriological Investigations and Diagnosis of Diphtheria, from May 4, 1893, to May 4, 1894." *Scientific Bulletin No. 1,* Health Department, City of New York. New York: Martin B. Brown Printer, 1895. Later published as "Diphtheria and Pseudo-Diphtheria" in *Medical Record,* 46 (1894): 385–401.

Park, William, and Charles Bolduan. "Mortality." In *The Bacteriology of Diphtheria,* ed. G. H. F. Nuttall and G. S. Graham-Smith. Cambridge: Cambridge University Press, 1913.

———. "The Value of Diphtheria Antitoxin in the Treatment of Diphtheria as Established by Ten Years Trial." *Annual Report of the Board of Health of the City of New York for the Year ending Dec. 31, 1905*, 2:515–61.

Park, William, M. C. Schroeder, and A. Zingher. "The Control of Diphtheria." *American Journal of Public Health* 13 (1923): 23–32.

Park, William, and Anna W. Williams. "The Production of Diphtheria Toxin." *Journal of Experimental Medicine* 1 (1896): 164–85.

———. *Who's Who Among the Microbes*. New York: The Century Co., 1929.

Park, William, A. Williams, and C. Krumweide. *Pathogenic Microorganisms*. Philadelphia: Lea & Febiger, 1924. (Eighth edition of Park's text originally titled *Bacteriology in Medicine and Surgery*.)

Park, William, and Abraham Zingher. "Practical Applications Obtained from the Schick Reaction." *Proceedings of the New York Pathological Society*, n.s., 14 (1914–1915): 151–58.

———. "Active Immunization with Diphtheria Toxin-Antitoxin—Second Paper." *Journal of the American Medical Association* 65 (25 December 1915): 2216–20.

———. "Immunity Results Obtained from the Use of Diphtheria Toxin-Antitoxin Mixtures and the Use of the Schick Test." *Medical Record*, 90 (1916): 741.

———. "Immunization of the Infant against Diphtheria." *Medical Record* 94 (1918): 615.

Park, William, Abraham Zingher, and Harry M. Serota, "Active Immunization in Diphtheria and Treatment by Toxin-Antitoxin." *Journal of the American Medical Association* 63 (5 September 1914): 859.

———. "Active Immunization with Diphtheria Toxin-Antitoxin—Second Paper," *Journal of the American Medical Association* 65 (25 December 1915): 2217.

———. "The Schick Reaction and Its Practical Applications." *Archives of Pediatrics* 31 (1914): 481.

"Preliminary Report of the Committee on Diphtheria Bacilli in Well Persons." *Journal of the Massachusetts Association of Boards Of Health* 11, no. 1 (1901): 1–15.

Prudden, Lillian E., ed. *Biographical Sketches and Letters of T. Mitchell Prudden*. New Haven: Yale University Press, 1927.

Prudden, T. Mitchell. "On the Etiology of Diphtheria: An Experimental Study." *American Journal of Medical Sciences*, n.s. 97 (April–May 1889): 329–50.

———. "On Koch's Methods of Studying Bacteria, Particularly Those Causing Asiatic Cholera." *8th Annual Report of the Board of Health of the State of Connecticut, for the Year ending November 1, 1885* (1886): 213–30.

———. "The Public Health: The Duty of the Nation in Guarding It." *Century Magazine* 46 (June 1893): 245–47.

———. *The Story of Bacteria*. New York: Putnam, 1889.

———. "Studies on the Etiology of Diphtheria." *Medical Record* 39 (18 April 1891): 445–50.

"Report of the American Pediatric Society Collective Investigation into the Use of Antitoxin in the Treatment of Diphtheria in Private Practice." *Archives of Pediatrics* 13 (July 1896): 481–503.

"Report of the Committee on the Cause and Prevention of Diphtheria." *Public Health, Papers and Reports,* American Public Health Association, 16 (1890): 97–109.

"Report of the Committee on Diphtheria Bacilli in Well Persons." *Journal of the Massachusetts Association of Boards of Health* 12 (1902): 74–95.

Roux, E., and L. Martin. "Contribution a l'étude de la diphthérie (serum therapie)." *Annales de L'Institut Pasteur* 6 (1894): 609.

Roux, E., L. Martin, and A. Chaillou. "Trois cents cas de diphthérie traites par le serum antidiphtérique." *Annales de L'Institut Pasteur* 8 (1894): 640.

Roux, E., and A. Yersin. "Contribution a l'étude de la diphtérie." *Annales de L'Institut Pasteur* 2 (1888): 629–61.

———. "Contribution a l'étude de la diphtérie." *Annales de L'Institut Pasteur* 3 (1889) 273–88.

Schick, Béla. "The Prevention and Control of Diphtheria." *Boston Medical and Surgical Journal* 188 (March 1, 1923): 255–58.

Schroeder, M. C. "The Duration of Active Immunity after Injections of Diphtheria Toxin-Antitoxin." *Medical Record,* 3 September 1921, 437.

———. "The Duration of the Immunity Conferred by the Use of Diphtheria Toxin-Antitoxin." *Archives of Pediatrics* 38 (1921): 369–72.

Smith, J. Lewis. "Membranous Croup; Diphtheritic Croup; True Croup." *Transactions of the New York State Medical Association* 1 (1884): 438.

Smith, Theobald. "Active Immunity Produced by So-called Balanced or Neutral Mixtures of Diphtheria Toxin and Antitoxin." *Journal of Experimental Medicine* 2 (March 1909): 241–56.

———. "The Conditions Which Influence the Appearance of Toxin in Cultures of the Diphtheria Bacillus." *Transactions of the Association of American Physicians* 9 (1896): 37.

———. "The Decline of Infectious Disease in Its Relation to Modern Medicine." *Journal of Preventive Medicine* 2 (1929): 345–63.

———. "The Degree and Duration of Passive Immunity to Diphtheria Antitoxin Transmitted by Immunized Female Guinea Pigs to Their Immediate Offspring." *Journal of Medical Research* 16 (1907): 359–70.

Solis-Cohen, Myer. "Diphtheria 'Carriers': Their Discovery and Control." *Journal of the American Medical Association* 52 (9 January 1909): 111–17.

Stella, Antonio. "The Effect of Urban Congestion on Italian Women and Children." *Medical Record* 74 (2 May 1908): 722–32.

Sterne, Simon. "Mayor Strong's Experiment in New York City." *Forum* 23 (1897): 539–53.

Thomson, W. H. "How the Facts About the Antitoxin Treatment of Diphtheria Should Be Estimated," *Medical Record*, 20 June 1896, 893.

Van Etten, Nathan B. "The Six Hundred and Fiftieth Man in Diphtheria Prevention," *New York State Journal of Medicine* 15 (April 1929): 452.

Vaughan, Victor [C.]. *Epidemiology and Public Health*. St. Louis: C. V. Mosby Co., 1922.

———. *A Doctor's Memories*. Indianapolis: Bobbs-Merrill, 1926.

Virchow, Rudolf. "Diphtheria and Croup." In *Collected Essays on Public Health and Epidemiology*, ed. L. J. Rather. Canton, Mass.: Science History Publications, 1985.

Von Pirquet, Clemens, and Béla Schick. *Serum Sickness*. Translated by Béla Schick. Baltimore: Williams and Wilkins, 1951.

Von Sholly, Anna I. "A Contribution to the Statistics on the Presence of Diphtheria Bacilli in Apparently Healthy Throats." *Journal of Infectious Disease* 4 (1907): 337–46.

Wald, Lillian. *The House on Henry Street*. New York: Henry Holt, 1915.

Welch, William [H.]. "Bacteriological Investigations of Diphtheria in the United States." *American Journal of Medical Sciences*, n.s. 107 (1894): 427–61.

———. *Papers and Addresses*. Baltimore: Johns Hopkins University Press, 1920.

Welch, William, and Alexander C. Abbott. "The Etiology of Diphtheria." *Bulletin of the Johns Hopkins Hospital* 2 (February–March 1891): 25–31.

Wesbrook, F. F., L. B. Wilson, and O. McDaniel. "Varieties of the Bacillus Diphtheriae." *Transactions of the Association of American Physicians*, 25 (1900): 198–223.

Williams, Anna [Wessels]. "The Morphology of the Diphtheria Bacillus." *Proceedings of the New York Pathological Society* 1 (1901): 93.

———. "Persistence of varieties of the bacillus diphtheriae and diphtheria-like bacilli." *Journal of Medical Research* 8, n.s. 5 (July–December 1902): 83–108.

Wilson, Charles, Walter Wyman, and Cyrus Edson. "Safeguards Against the Cholera." *North American Review* 331 (October 1892): 485–503.

Winslow, C.-E. A. *The Evolution and Significance of the Modern Public Health Campaign*. New Haven, Conn.: Yale University Press, 1923.

———. *The Life of Hermann M. Biggs*. Philadelphia: Lea & Febiger, 1929.

Winslow, C.-E. A., and S. Zimand. *Health Under the El: The Story of the Bellevue-Yorkville Health Demonstration Project in Mid-Town New York*. New York: Harper and Brothers, 1937.

Winters, Joseph E. "Clinical Observations Upon the Use of Antitoxin in Diphtheris:

A Report of Personal Investigation of this Treatment in the Principal Fever Hospitals of Europe During the Summer of 1895." *Medical Record* 49 (20 June 1895): 877–95.

——. *Diphtheria and Its Management.* New York: Trow's Printing and Bookbinding, 1885.

——. "Diphtheria and Its Management: Are Membranous Croup and Diphtheria Distinct Diseases?" *Medical Record* 28 (5 December 1885): 617–26.

Wright, Louis T. "The Schick Test, with Especial Reference to the Negro." *Journal of Infectious Diseases* 21 (1917): 265–68.

Zingher, Abraham. "Active Immunization of Infants Against Diphtheria." *American Journal of Diseases of Children* 16 (1918): 83–102.

——. "Diphtheria Prevention Among Children of Pre-school Age." *Journal of the American Medical Association* 80 (1923): 456–60.

——. "Diphtheria Prevention Work in the Public Schools of New York City." *Journal of the American Medical Association* 77, no. 11 (1921): 835–43.

——. "Diphtheria Preventive Work in the Public Schools of New York City." *Medical Record*, 3 September 1921, 438.

——. "Interpretation of the Schick Reaction in Recruits for the National Army." *Journal of the American Medical Association* 70, no. 4 (16 January 1918): 227–28.

——. "Methods of Using Diphtheria Toxin in the Schick Test and Controlling the Reaction." *American Journal of Diseases of Children* 11 (1916): 269–75.

——. "The Schick Test Performed on more than 150,000 Schoolchildren in Public and Parochial Schools in New York (Manhattan and the Bronx)." *American Journal of Diseases of Children* 25 (1923): 392–405.

Secondary Sources from the Published Literature

Ackerknecht, Erwin H. *Medicine in the Paris Hospital.* Baltimore: Johns Hopkins University Press, 1987.

——. *Rudolph Virchow: Doctor, Statesman, Anthropologist.* Madison: University of Wisconsin Press, 1953.

Amsterdamska, Olga. "Medical and Biological Constraints: Early Research on Variation in Bacteriology." *Social Studies of Science*, 17 (1987): 657–87.

Apple, Rima. *Mothers and Medicine: A Social History of Infant Feeding, 1890–1950.* Madison: University of Wisconsin Press, 1987.

Bisgard, H. M., et al. "Virtual Elimination of Respiratory Diphtheria in the United

States." *Abstracts of the 36th Interscience Conference on Antimicrobials and Chemotherapy, September 15–18, 1996, New Orleans, La.* Abstract No. K166:280.

Bloomfield, Arthur. *A Bibliography of Internal Medicine: Communicable Diseases.* Chicago: University of Chicago Press, 1958.

Bonner, Thomas N. *American Doctors and German Universities.* Lincoln: University of Nebraska Press, 1963.

Brandt, Allan. "AIDS: From Social History to Social Policy." In *AIDS: The Burdens of History*, ed. Elizabeth Fee and Daniel M. Fox. Berkeley: University of California Press, 1988.

Buck, Peter S. "Order and Control: The Scientific Method in China and the United States." *Social Studies of Science* 5 (1975): 237–67.

Bulloch, William. *The History of Bacteriology.* New York: Dover Publications, 1938.

Burrows, W. *Textbook of Microbiology.* Philadelphia: W. B. Saunders, 1979.

Carter, K. Codell. "Koch's Postulates in Relation to the Work of Jacob Henle." *Medical History* 29 (1985): 353–74.

Cassedy, James. *Charles V. Chapin and the Public Health Movement.* Cambridge, Mass.: Harvard University Press, 1962.

Cassel, John. "The Contribution of the Social Environment to Host Resistance." *American Journal of Epidemiology* 102 (1976): 107–23.

Caulfield, Ernest. "A History of the Terrible Epidemic, Vulgarly Called the Throat Distemper, As It Occurred in His Majesty's New England Colonies Between 1735 and 1740." *Yale Journal of Biology and Medicine* 11 (1938): 219–35.

Coleman, William. *Yellow Fever in the North: The Methods of Early Epidemiology.* Madison: University of Wisconsin Press, 1987.

Condran, Gretchen A. "Changing Patterns of Epidemic Disease in New York City." In *Hives of Sickness: Public Health and Epidemics in New York City*, 27–41. New Brunswick, N.J.: Rutgers University Press, 1995.

Condran, Gretchen A., and Eileen Crimmins-Gardner. "Public Health Measures and Mortality in U.S. Cities in the Late Nineteenth Century." *Human Ecology* 6 (1978): 27–54.

Condran, Gretchen A., and Ellen Kramarow. "Child Mortality Among Jewish Immigrants to the United States." *Journal of Interdisciplinary History* 22 (Autumn 1991): 250–53.

Cooter, Roger. "Anticontagionism and History's Medical Record." In *The Problem of Medical Knowledge*, ed. P. Wright and A. Treacher. Edinburgh: Edinburgh University Press, 1983.

Cunningham, Andrew. "Transforming Plague: The Laboratory and the Identity of Infectious Disease." In *The Laboratory Revolution in Medicine*, ed. Andrew Cunningham and Perry Williams, 209–45. Cambridge: Cambridge University Press, 1992.

Davis, B., et al., *Microbiology*, 4th ed. Philadelphia: J. B. Lippincott Co., 1990.

Davis, Michael M. *Immigrant Health and the Community*. New York: Harper and Brothers, 1921; Montclair, N.J.: Patterson Smith Publishing, 1971.

De Kruif, Paul. *Microbe Hunters*. New York: Harcourt, Brace and Co., 1926.

Delaporte, Francois. "The History of Medicine According to Foucault." In *Foucault and the Writing of History*, ed. J. Goldstein, pp. 137–49. Oxford: Basil Blackwell, 1994.

Dixon, Bernard. "Infectious Diseases—A 'Dead' Specialty Revives." *Bio/Technology* 9 (September 1991): 779.

Dowling, Harry F. *Fighting Infection: Conquests of the Twentieth Century*. Cambridge, Mass.: Harvard University Press, 1977.

Duffy, John. "The American Medical Profession and Public Health: From Support to Ambivalence." *Bulletin of the History of Medicine* 53 (Spring 1979): 1–15.

——. *A History of Public Health in New York City, 1866–1966*. New York: Russell Sage Foundation, 1974.

——. *The Sanitarians: A History of American Public Health*. Urbana: University of Illinois Press, 1992.

Dwork, Deborah. "Health Conditions of Immigrant Jews on the Lower East Side of New York: 1880–1914." *Medical History* 25 (1981): 1–40.

Eggerth, Arnold. *The History of the Hoagland Laboratory*. Brooklyn, N.Y.: 1960. [s.n.]

English, Peter C. "Diphtheria and Theories of Infectious Disease: Centennial Appreciation of the Critical Role of Diphtheria in the History of Medicine." *Pediatrics* 76 (1985): 1–9.

Evans, Richard J. *Death in Hamburg: Society and Politics in the Cholera Years, 1830–1910*. Oxford: Clarendon Press, 1987.

Faber, Knud. *Nosography in Modern Internal Medicine*. New York: Paul B. Hoeber, Inc., 1923.

Fee, Elizabeth, and Daniel M. Fox, eds. *AIDS: The Burdens of History*. Berkeley: University of California Press, 1988.

——. *AIDS: The Making of a Chronic Disease*. Berkeley: University of California Press, 1992.

Fleming, Donald. *William H. Welch and the Rise of Modern Medicine*. Baltimore: Johns Hopkins University Press, 1954.

Fox, Daniel. "Social Policy and City Politics: Tuberculosis Reporting in New York City, 1889–1900." In *In Sickness and Health in America*, 2d ed., ed. Judith Walzer

Leavitt and Ronald L. Number, 415–31. Madison: University of Wisconsin Press, 1985.

Freudenberg, Nicholas. "AIDS Prevention in the United States: Lessons from the First Decade." *International Journal of Health Services* 20, no. 4 (1990): 589–99.

Fujimura, Joan. "Crafting Science: Standardized Packages, Boundary Objects, and 'Translation.'" In *Science as Practice and Culture,* ed. A. Pickering, 169–70. Chicago: University of Chicago Press, 1992.

Gordon, Linda. "Family Violence, Feminism, and Social Control." In *Unequal Sisters: A Multi-Cultural Reader in U.S. Women's History,* ed. E. DuBois and V. Ruiz, 146–47. New York: Routledge, 1990. Originally published in *Feminist Studies,* 12 (Fall 1986).

Greenough, Paul. "Global Immunization and Culture: Compliance and Resistance in Large-Scale Public Health Campaigns." *Social Science and Medicine,* 41 (1995): 605–7.

Gronowicz, Antoni. *Béla Schick and the World of Children.* New York: Abelard-Schuman, 1954.

Hamilton, Diane. "The Cost of Caring: The Metropolitan Life Insurance Company Visiting Nurse Service, 1909–1953." *Bulletin of the History of Medicine* 63 (1989): 414–34.

Hammonds, Evelynn M. "Race, Sex, AIDS: The Construction of 'Other.'" *Radical America* 20, no. 6 (1987): 28–36.

———. "Missing Persons: African-American Women, AIDS, and the History of Disease," *Radical America* 24, no. 2 (April–June 1990): 7–24.

Hardy, Anne. *The Epidemic Streets: Infectious Disease and the Rise of Preventive Medicine.* Oxford: Clarendon Press, 1993.

Hofstadter, Richard. *The Progressive Movement, 1900–1915.* Englewood Cliffs, N.J.: Prentice-Hall, 1963.

Jardine, Nicholas. "The Laboratory Revolution in Medicine as Rhetorical and Aesthetic Accomplishment." In *The Laboratory Revolution in Medicine,* ed. Andrew Cunningham and Perry Williams, 304–23. Cambridge: Cambridge University Press, 1992.

Katz, Michael. *In the Shadow of the Poorhouse.* New York: Basic Books, 1986.

Kleinman, Lawrence. "To End an Epidemic: Lessons from the History of Diphtheria." *New England Journal of Medicine* 326 (1992): 773–77.

Kramer, Howard D. "The Germ Theory and the Early Public Health Program in the United States." *Bulletin of the History of Medicine* 22 (1948): 233–47.

Kraut, Alan. "East Side Parents Storm the Schools: Public Schools and Public

Health." In *Silent Travelers: Germs, Genes and the "Immigrant Menace,"* 226–56. New York: Basic Books, 1994.

———. "Silent Travelers: Germs, Genes, and American Efficiency, 1890–1924." *Social Science History* 12 (1988): 377–94.

Ladd-Taylor, Molly. "Hull House Goes to Washington: Women and the Children's Bureau." In *Gender, Class, Race and Reform in the Progressive Era*, ed. Noralee Frankel and Nancy S. Dye. Lexington: University of Kentucky Press, 1991.

Latour, Bruno. *The Pasteurization of France.* Translated by Alan Sheridan and John Law. Cambridge, Mass.: Harvard University Press, 1988.

Leavitt, Judith W. " 'Typhoid Mary' Strikes Back: Bacteriological Theory and Practice in Early Twentieth-century Public Health." *Isis*, 83 (1992): 608–29.

———. Letter to the editor. *Isis*, 86, no. 4 (1995): 617–18.

Lechevalier, Hubert A., and Morris Solotorovsky. *Three Centuries of Microbiology.* New York: McGraw-Hill, 1965.

Lederer, Susan. *Subjected to Science: Human Experimentation in America Before the Second World War.* Baltimore: Johns Hopkins University Press, 1995.

Ledingham, J., and J. A. Arkwright. *The Carrier Problem in Infectious Diseases.* London: Edward Arnold, 1912.

Lewis, Jane. "The Prevention of Diphtheria in Canada and Britain 1914–1945." *Journal of Social History* 20 (1986): 163–76.

Lewistein, Bruce. "Industrial Life Insurance, Public Health Campaigns, and Public Communication of Science, 1908–1951," *Public Understanding of Science,* 1 (1992): 347–65.

Liebenau, Jonathan. "Public Health and the Production of Diphtheria Antitoxin in Philadelphia." *Bulletin of the History of Medicine* 61 (1987): 216–36.

Maulitz, Russell. "Rudolf Virchow, Julius Cohnheim and the Program of Pathology." *Bulletin of the History of Medicine* 52 (1982): 162–82.

———. "Physician vs. Bacteriologist: The Ideology of Science in Clinical Medicine." In *The Therapeutic Revolution: Essays in the Social History of American Medicine,* ed. Morris Vogel and Charles Rosenberg, pp. 91–107. Philadelphia: University of Pennsylvania Press, 1979.

Mazumdar, Pauline. "Immunity in 1890." *Journal of the History of Medicine* 27 (1972): 312–24.

McFarland, Joseph. "The Beginnings of Bacteriology in Philadelphia." *Bulletin of the History of Medicine* 5 (1937): 149–98.

Meckel, Richard. *Save the Babies: American Public Health Reform and the Prevention of Infant Mortality 1850–1929.* Baltimore: Johns Hopkins University Press, 1990.

Mendelsohn, J. Andrew. "'Typhoid Mary' Strikes Again: The Social and the Scientific in the Making of Modern Public Health." *Isis*, 86 (1995): 268–77.

Mettler, Cecilia. *History of Medicine*, ed. Fred A. Mettler. Philadelphia: Blakiston Co., 1947.

Morantz, Regina Markell. "Feminism, Professionalism and Germs: The Thought of Mary Putnam Jacobi and Elizabeth Blackwell." *American Quarterly* 34 (1982): 459–78.

Mormon, Edward T. "Clinical Pathology in America, 1865–1915: Philadelphia as a Test Case." *Bulletin of the History of Medicine* 58 (1984): 198–214.

Nelkin, Dorothy. "AIDS and the News Media." *Milbank Quarterly* 69 (1991): 293–307.

Parish, H. J. *Victory with Vaccines: The Story of Immunization*. Edinburgh: E. & S. Livingstone, 1968.

Pelling, Margaret. *Cholera, Fever and English Medicine, 1825–1865*. Oxford: Oxford University Press, 1978.

Pernick, Martin, "Thomas Edison's TB Films: Mass Media and Health Propaganda," *Hastings Center* 8 (1978): 21–27.

Pickstone, John V., ed. *Medical Innovations in Historical Perspective*. New York: St. Martin's Press, 1992.

Popovic, T., M. Wharton, et al. "Are We Ready for Diphtheria? A Report from the Diphtheria Diagnostic Workshop, Atlanta, 11 and 12 July 1994." *Journal of Infectious Diseases* 171 (1995): 765–67.

Preston, Samuel, and Michael Haines. *Fatal Years: Child Mortality in Late 19th Century America*. Princeton, N.J.: Princeton University Press, 1991.

Quiroga, Virginia. "Diphtheria and Medical Therapy in Late 19th Century New York City." *New York State Journal of Medicine* 90, no. 5 (May 1990): 256–62.

Ravenal, M. P. *A Half Century of Public Health*. New York: American Public Health Association, 1921.

Reiser, Stanley J. *Medicine in the Reign of Technology*. New York: Cambridge University Press, 1978.

Richmond, Phyllis Allen. "American Attitudes Toward the Germ Theory of Disease (1860–1880)." *Journal of the History of Medicine* 9 (1954): 428–54.

Riis, Jacob. *How the Other Half Lives: Studies Among the Tenements of New York*. New York: Dover Publications, 1977. Unabridged reprint of the text of the 1890 edition.

Robinton, Elizabeth. "Anna Wessels Williams." In *Notable American Women: The Modern Period*, ed. B. Sicherman and C. Green, 737–39. Cambridge, Mass.: Harvard University Press, 1980.

Rogers, Naomi. *Dirt and Disease: Polio Before FDR.* New Brunswick: Rutgers University Press, 1992.

Rosen, George. "The First Neighborhood Health Center Movement—Its Rise and Fall." *American Journal of Public Health* 61 (1971): 195–99.

———. *A History of Public Health*, expanded ed. Baltimore: Johns Hopkins University Press, 1993.

Rosenberg, Charles. *The Care of Strangers: The Rise of America's Hospital System.* New York: Basic Books, 1987.

———. *The Cholera Years: The United States in 1832, 1849 and 1866.* Chicago: University of Chicago Press, 1987.

———. "Disease in History: Frames and Framers," *Milbank Quarterly* 67 (suppl. 1, 1989): 1–15.

———. "Framing Disease: Illness, Society and History." In *Framing Disease: Studies in Cultural History,* ed. Charles Rosenberg and Janet Golden, xiii–1. New Brunswick: Rutgers University Press, 1992.

———. "Making It in Urban Medicine: A Career in the Age of Scientific Medicine." *Bulletin of the History of Medicine* 64 (1990): 163–86.

———. "The Practice of Medicine in New York a Century Ago." *Bulletin of the History of Medicine* 41 (1967): 233–53.

Rosenkrantz, Barbara Gutmann. *Public Health and the State: Changing Views in Massachusetts, 1842–1936.* Cambridge, Mass.: Harvard University Press, 1972.

———. "Cart Before Horse: Theory, Practice and Professional Image in American Public Health, 1870–1920." *Journal of the History of Medicine and Allied Sciences* 29 (1974): 55–73.

Rosner, David J., ed., *Hives of Sickness: Public Health and Epidemics in New York City.* New Brunswick, N.J.: Rutgers University Press, 1995.

Rothman, David J. "The State as Parent: Social Policy in the Progressive Era." In *Doing Good: The Limits of Benevolence,* ed. Willard Gaylin, Ira Glasser, Steven Marcus, and David Rothman. New York: Pantheon Books, 1978.

Rothstein, William G. *American Physicians in the Nineteenth Century: From Sects to Science.* Baltimore: Johns Hopkins University Press, 1972.

Schaeffer, Morris. "William H. Park (1863–1939): His Laboratory and His Legacy." *American Journal of Public Health* 75 (1985): 1296–302.

Schaffer, Simon. "The Eighteenth Brumaire of Bruno Latour." *Studies in the History and Philosophy of Science,* 22, no. 1 (1991): 175–92.

Schoener, Allan. *Portal to America.* New York: Holt, Rinehart and Winston, 1967.

Schudson, Michael. *Discovering the News.* New York: Basic Books, 1978.

Solberg, Sheryl Gay. "AIDS Drugs Give Little Hope to Thousands Unable to Pay." *New York Times,* 14 October 1997, p. 1, A22.

Sontag, Susan. *AIDS and Its Metaphors*. New York: Farrar, Strauss and Giroux, 1989.

Starr, Paul. *The Social Transformation of American Medicine*. New York: Basic Books, 1982.

Starr, S. L., and J. R. Griesemer. "Institutional Ecology, 'Translations,' and Boundary Objects: Amateurs and Professionals in Berkeley's Museum of Vertebrate Zoology, 1907–39." *Social Studies of Science*, 19 (1989): 387–420.

Stevenson, Lloyd. "Science Down the Drain: On the Hostility of Certain Sanitarians to Animal Experimentation, Bacteriology, and Immunology." *Bulletin of the History of Medicine*, 29 (1955): 1–26.

Truax, Rhoda. *The Doctors Jacobi*. Boston: Little, Brown and Co., 1952.

Van Ingen, Philip. *The New York Academy of Medicine: Its First Hundred Years*. New York: Columbia University Press, 1949.

Vogel, Morris, and Charles Rosenberg, eds. *The Therapeutic Revolution*. Philadelphia: University of Pennsylvania Press, 1979.

Walsh, James J. *History of Medicine in New York: Three Centuries of Medical Progress*. New York: National Americana Society, 1919.

Ward, David. "The Internal Spatial Differentiation of Immigrant Residential Districts." In *Geographic Perspectives on America's Past*, ed. David Ward, 335–43. New York: Oxford University Press, 1979.

———. "The Internal and Spatial Structure of Immigrant Residential Districts in the Late Nineteenth Century." *Geographical Analysis* 4, no. 1 (1969): 357–73.

———. *Poverty, Ethnicity and the American City, 1890–1925: Changing Concepts of the Slum and Ghetto*. Cambridge: Cambridge University Press, 1989.

Warner, John Harley. "The Fall and Rise of Professional Mystery: Epistemology, Authority, and the Emergence of Laboratory Medicine in Nineteenth-Century America." In *The Laboratory Revolution in Medicine*, ed. Andrew Cunningham and Perry Williams, 110–141. Cambridge: Cambridge University Press, 1992.

———. "Ideals of Science and Their Discontents in Late Nineteenth-Century American Medicine." *Isis* 82 (1991): 454–78.

———. *The Therapeutic Perspective: Medical Practice, Knowledge, and Identity in America, 1820–1885*. Cambridge, Mass.: Harvard University Press, 1986.

Weindling, Paul. "From Isolation to Therapy: Children's Hospitals and Diphtheria in Fin de Siècle Paris, London and Berlin." In *In the Name of the Child: Health and Welfare, 1880–1940*, ed. Roger Cooter, 124–45. London: Routledge, 1992.

———. "From Medical Research to Clinical Practice." In *Medical Innovations in Historical Perspective*, ed. John V. Pickstone, 72–83. New York: St. Martin's Press, 1992.

Williams, Phyllis. "The South Italians: Health and Hospitals." In *The Italians: Social*

Backgrounds of An American Group, ed. Francesco Cordasco and Eugene Bucchi-
oni, 457–75. Clifton, N. J.: Augustus M. Kelley, Pub., 1974.

Wilson, G. S., A. A. Miles, and M. T. Parker. *Topley and Wilson's Principals of Bacteriology, Virology and Immunity.* Baltimore: Williams and Wilkins, 1984.

Ziporyn, Terra. *Disease in the Popular American Press: The Case of Diphtheria, Typhoid Fever and Syphilis, 1870–1920.* New York: Greenwood Press, 1988.

INDEX

LIBRARY OF CONGRESS CATALOGING-IN-PUBLICATION DATA

Hammonds, Evelynn Maxine.
    Childhood's deadly scourge : the campaign to control diphtheria
in New York City, 1880–1930 / Evelynn Maxine Hammonds.
        p.   cm.
    Includes bibliographical references and index.
    ISBN 0-8018-5978-6 (alk. paper)
        1. Diphtheria — New York (State) — New York — History — 19th century.
    2. Diphtheria — New York (State) — New York — History — 20th century.
    I. Title.
    RA644.D6H36   1999
    614.5′123′097471 — dc21   98-38209
                                                                            CIP